SpringerBriefs in Applied Sciences and Technology

For further volumes:
http://www.springer.com/series/8884

Alexei Kryukov · Vladimir Levashov
Yulia Puzina

Non-Equilibrium Phenomena Near Vapor–Liquid Interfaces

 Springer

Prof. Alexei Kryukov
Dr. Vladimir Levashov
Dr. Yulia Puzina
Moscow Power Engineering Institute
Moscow
Russia

ISSN 2191-530X   ISSN 2191-5318 (electronic)
ISBN 978-3-319-00082-4   ISBN 978-3-319-00083-1 (eBook)
DOI 10.1007/978-3-319-00083-1
Springer Heidelberg New York Dordrecht London

Library of Congress Control Number: 2013933571

Printed on acid-free paper

Springer is part of Springer Science+Business Media (www.springer.com)

Preface

It is known that in fluid dynamics boundary conditions are of huge importance, because they have relevance to heat and mass transfer efficiency across interface surfaces. Especially, in cases where the boundary condition is for the interface of a vapor or gas–vapor mixture and its liquid phase some difficult problems appear, which are not yet made clear. This is because the derivation of the boundary condition requires detailed information of molecular phenomena at the interface, whereas the governing equations (the Navier–Stokes equations in fluid dynamics) can be derived from macroscopic conservation laws. Thus, there is great necessity to formulate correctly the discontinuity of initial and boundary conditions for precise and productive solutions of different interface transfer problems. The joint solution of the Boltzmann kinetic and fluid dynamic equations is used for the study. On the base of this approach, the limiting matter possibilities in respect of transfer processes can be found. Another task is the study of the conjugate problem: vapor (vapor–gas) mixture–liquid (condensate) at condensation, evaporation, and boiling. The processes in following systems are investigated: superfluid helium (He-II), because efficiency of heat transfer in He-II is highest among all liquids, and liquid metals, because thermal conductivity of these substances is very high. The application of above mentioned approach for determination of the shape of interface surface at non-equilibrium conditions in film boiling is presented.

Contents

1 Introduction 1

2 Background for Pure (One Component) Substance 3
   References 7

3 Evaporation and Condensation of Vapor–Gas Mixtures 9
   3.1 Liquid Evaporation in Vapor–Gas Mixtures 10
       3.1.1 Problem and Solution Method 10
       3.1.2 Results and Analysis 12
   3.2 Condensation from Vapor–Gas Mixture 18
       3.2.1 Statement of the Problem 19
       3.2.2 Results and Discussion 21
   References 22

4 Motion of Vapor–Liquid Interfaces 25
   4.1 Motion of Helium II Bridges in Capillary Channels with Vapor
      at the Presence of Longitudinal Heat Flux 25
       4.1.1 Statement of the Problem 26
       4.1.2 Description of the Model, Basic Equations
           and Results 26
   4.2 Evolution of Vapor Films at Boiling of He-II
      and Ordinary Liquids 32
       4.2.1 Statement of the Problem and Basic Equations 32
       4.2.2 Results and Discussions 36
       4.2.3 The Model with Constant Interphase Temperature 36
       4.2.4 The Model Accounting for Heat Transfer
           in Superfluid Helium 38
   4.3 Conclusion 38
   References 39

5 Liquid–Vapor Interface Form Determination 41
   5.1 Statement of the Problem and Model 42
   5.2 Comparison with Experimental Data 45

5.3   Results Discussion   46
       5.3.1   Immersion Depth Dependence on Drop Mass   48
       5.3.2   Vapor Film Thickness Dependence on Time   49
       5.3.3   Liquid Properties Influence   49
5.4   Conclusion   50
References   51

6   **Summary**   53

Symbols

j_E	Mass flux density in evaporation process
j_C	Mass flux density in condensation process
T_S	Interface temperature
T	Temperature
T_∞	Temperature away from evaporation surface
R	Individual gas constant for vapor
l	Mean free path of vapor molecules
L	Typical size of system
$Kn = l/L$	Knudsen number
u	Average velocity
P	Pressure
λ_{CO_2}	Thermal conductivity of CO_2
r	Latent heat of de-sublimation per mass unit
n_v	Numerical density of vapor molecules
f	Velocity distribution function
x, y, z	Cartezian coordinate
t	Time
J_{ij}	Collision integrals
J^{diff}	Diffusion mass flux
D	Diffusion coefficient
$C_g = \rho_g/\rho$	Relative densities of "g" component
h_v, h_g	Enthalpy of component
e	Internal mixture energy
m_v and m_g	Molecular masses of "v" and "g" component accordingly
M_∞	Mach number
d	Diameter
g	Acceleration of gravity
K	Curvature
h	Vertical distance
L	Heat of evaporation
l	Length of He-II column
q	Heat flux density

R_w	Heater radius,
R_1	Vapor film radius, m
S	Entropy
s	Curve length
V, w	Velocity

Greek Symbols

β_E	Evaporation coefficient
β_C	Condensation coefficient
ρ_s	Saturated vapor density corresponding interface temperature T_s
ρ_∞	Vapour density away from evaporation surface
μ	Mixture viscosity
δ	Vapor film thickness
η	Dynamic viscosity
λ	Thermal conductivity
ν	Kinematic viscosity
ρ	Density
σ	Surface tension
ξ	Coefficient of hydraulic resistance
φ	Angle

Subscripts and Superscripts

E	Evaporation
C	Condensation
0	Bottom point
V	Vapour
g	Gas
$1, i$	Parameter of vapor–liquid interface
b	Far away from interface
s	Parameter of the saturation line
w	Parameter of heater
∞	Ambient
$'$	Parameter of liquid
$''$	Parameter of vapor

Chapter 1
Introduction

Development of continuum mechanics (fluid dynamics) methods in last century has provided progress of transfer phenomena description and calculation in bulk of solid, liquid and gas (vapor). However, it is well known that for successful solution of fluid dynamics equations corresponding initial and boundary conditions should be formulated. These conditions should be prescribed for all fluid dynamics values. Type of such prescription depends on concrete task. Nevertheless for correct solution of different heat and mass transfer problem there is necessity to take into account non-equilibrium near interfaces even in case when these surfaces are impenetrable. At the presence of evaporation and condensation, adsorption and desorption on the interfaces this necessity becomes more strong. The background of the problem with the emphasis on the previously suggested kinetic models for pure (one component) substance is presented in Chap. 2. In Chap. 3 the evaporation and condensation of vapor–gas mixtures research is given. Motion of vapor–liquid interfaces is considered and discussed in Chap. 4. The approach to the problem of vapor–liquid interfaces forms determination and corresponding calculations results are presented in Chap. 5. The main conclusions of the study are summarized in Chap. 6.

A. Kryukov et al., *Non-Equilibrium Phenomena Near Vapor–Liquid Interfaces*, SpringerBriefs in Applied Sciences and Technology, DOI: 10.1007/978-3-319-00083-1_1, © The Author(s) 2013

Chapter 2
Background for Pure (One Component) Substance

One way to develop accurate boundary conditions is the application of molecular-kinetic theory to creation of these conditions. From point of this theory view intensity of evaporation and condensation processes (mass flux density j) can be defined. Brief history of study development about mass flux density j determination at evaporation and condensation of pure vapor is following.

First correlation, which should be mentioned, is famous Herts-Knudsen (HK) formula attributed to Herts [1] and Knudsen [2]. Possible version of this formula at $\beta_E = \beta_C = \beta$, where β_E is the evaporation coefficient, β_C is the condensation coefficient:

$$ j = \beta \left[\rho_S \sqrt{RT_S/2\pi} - \rho_\infty \sqrt{RT_\infty/2\pi} \right] \tag{2.1} $$

Evaporation coefficient is defined as ratio of vapor mass flux density in evaporation process near the liquid (solid)—vapor interface j_E to mass flux density calculated in accordance with positive half of Maxwell distribution function with temperature of interface T_S and equilibrium (saturated vapor) density ρ_S corresponding to this temperature $\rho_S \sqrt{RT_s/2\pi}$. Thus $\beta_E = j_E / \rho_S \sqrt{RT_s/2\pi}$. Condensation coefficient β_C is defined as ratio of mass flux density of condensing on the interface molecules j_C to mass flux density of whole flux of molecules striking with interface j_-: $\beta_C = j_C / j_-$.

HK formula is based on the assumption that as emitted (evaporated) from interface molecules and as molecules moving to this surface are described by half Maxwellian velocity distributions with constant in any point of space occupied by vapor ρ_s, T_s, ρ_∞, T_∞. Schematically this distribution function is illustrated by Fig. 2.1.

Hence relationship (2.1) is valid only for free-molecular flow that is at $Kn \to \infty$ where Kn is Knudsen number—ratio of mean free path of vapor molecules l to typical system size L. Thus Herts-Knudsen formula cannot be used for determination of mass flux density at $Kn \to 0$.

In continuum limit at $Kn \to 0$ velocity distribution function for vapor far (in the scale of mean free path) from interface cannot be presented by two-side

A. Kryukov et al., *Non-Equilibrium Phenomena Near Vapor–Liquid Interfaces*, SpringerBriefs in Applied Sciences and Technology, DOI: 10.1007/978-3-319-00083-1_2, © The Author(s) 2013

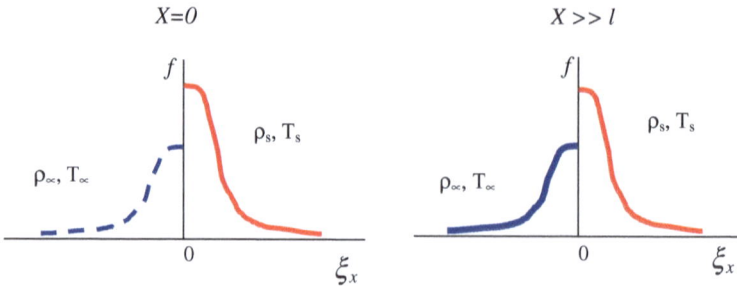

Fig. 2.1 Half-Maxwellians

Maxwellian like in HK formula because there is the motion of bulk vapor and corresponding distribution function looks like in Fig. 2.2.

First attempts to take into account velocity of bulk vapor, directed away from interface, were made in papers [3–5]. Crout was first who used a distribution function for the vapor remote from the interface which would give the bulk vapor velocity u_∞. However, Schrage in [5] has criticized such application of ellipsoidal Crout's distribution because this function is incompatible with distribution of molecules evaporated from the interface. R. W. Schrage has proposed two types of distribution function for molecules moving to the interface near it. In the first of them this distribution function is just part of distribution far from interface with same bulk velocity, temperature and density for molecules moving to the interface $u_\infty, \rho_\infty, T_\infty$.

Kucherov and Rikenglaz [6] have applied thirteen-moment distribution for study of weak evaporation and condensation at which vapor bulk velocity is much smaller the vapor sonic velocity. Due to this strong inequality the description has been simplified considerably, became linear and distribution function has been transformed in shifted Maxwellian distribution like in Schrage's approach. As a result the authors have deduced evident formula for determination of mass flux density in this process at $\beta_E = \beta_C = \beta = 1$ in continuum limit. Calculation in accordance with this formula was in two times more value obtained from Herts-Knudsen correlation

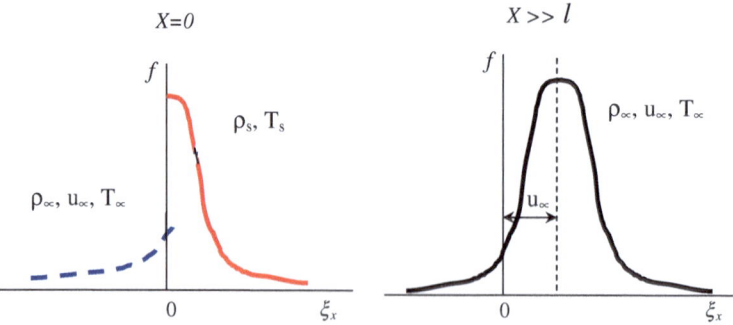

Fig. 2.2 Distribution functions

(2.1). Same result can be derived from [5] where relationship j, ρ_s, T_s, ρ_∞, T_∞ is presented but in implicit form concerning j. A little later in paper [7] Kucherov and Rikenglaz have made generalization for arbitrary evaporation–condensation coefficient at the assumption $\beta_E = \beta_C = \beta$ and have presented the following formula:

$$j = \frac{\beta}{1 - 0.5\beta}\left[\rho_s\sqrt{RT_S/2\pi} - \rho_\infty\sqrt{RT_\infty/2\pi}\right] \qquad (2.2)$$

For further analysis we need the generalization (2.2) for the case $\beta_E \neq \beta_C$. In order to derive this correlation let us to repeat deduction in Kucherov's and Rikenglaz's manner but taking into account that $\beta_E \neq \beta_C$. This procedure can be done for linearized statement (weak evaporation) that is at $V_\infty/\sqrt{2RT_\infty} < <1$, $\rho_s \approx \rho_\infty$, $T_s \approx T_\infty$. Resulting correlation is following:

$$j = \frac{1}{1 - 0.5\beta_C}\sqrt{RT/2\pi}[\beta_E\rho_s - \beta_C\rho_\infty] \qquad (2.3)$$

where $T \approx T_s \approx T_\infty$. At $\beta_E = \beta_C = \beta$ formula (2.3) transforms in (2.2). For weak evaporation and condensation $P_S = \rho_s RT_S \approx \rho_s RT$ and $P_\infty = \rho_\infty RT_\infty \approx \rho_\infty RT$. Hence (2.3) can be presented as:

$$j = \frac{1}{1 - 0.5\beta_C}\frac{[\beta_E P_S - \beta_C P_\infty]}{\sqrt{2\pi RT}} \qquad (2.3a)$$

Relationships (2.2) and (2.3) have been obtained on the base only conservation equations and known prescribed distribution function for molecules moving to the interface. Labuntsov [8], Muratova and Labuntsov [9] have solved the Boltzmann kinetic equation for weak evaporation and condensation and deduced from these solutions more accurate formula for $\beta_E = \beta_C = \beta$ instead (2.2):

$$j = \frac{\beta}{1 - 0.4\beta}\frac{(P_S - P_\infty)}{\sqrt{2\pi RT}} \qquad (2.4)$$

Comparison (2.2) with (2.4) shows that at $\beta_E \neq \beta_C$ instead (2.3a) more accurate correlation should be used:

$$j = \frac{1}{1 - 0.4\beta_C}\frac{(\beta_E P_S - \beta_C P_\infty)}{\sqrt{2\pi RT}} \qquad (2.4a)$$

Correlation (2.2) and (2.4) above are valid only for weak processes when bulk velocity of vapor flow is much smaller sound velocity. At the rising evaporation or condensation intensity non-equilibrium of vapor increases and distribution function changes more strongly. The good enough solutions of this problem on the base of the Boltzmann kinetic equation and its models for high evaporation intensity have been studied and were presented in different papers [10–19]. For calculation of mass flux density in evaporation problem in paper [15] the following formula were suggested for $\beta = 1$:

$$j = 0.6\sqrt{2RT_S}(\rho_S - \rho_\infty)\sqrt{\frac{\rho_\infty}{\rho_S}} \tag{2.5}$$

For calculation of subsonic strong condensation the correlation below were obtained on the base corresponding molecular-kinetic approaches:

$$j = 1.67\frac{P - P_S}{\sqrt{2\pi RT_\infty}} \cdot \left\{1 + 0.51 \cdot \ln\left[\frac{P\sqrt{T_\infty}}{P_S\sqrt{T_S}}\right]\right\} \tag{2.6}$$

As example the comparison of application of equilibrium and non-equilibrium boundary conditions to condensation in solid state (de-sublimation) transfer process is presented bellow.

Condensation of gas carbon dioxide (CO_2) on a flat surface is studied in steady statement. Temperature of this surface T_w is known and constant during condensation process (see Fig. 2.3).

Gas (vapor) parameters far from the interface in comparison with of gas molecules mean free path are considered as known. It is supposed that on the plate surface layer of solid carbon dioxide exists. Thickness of this deposit δ is known and does not change during the whole process. Four unknown values are needed to determine: temperature of condensate interface—T_s, vapor pressure corresponding to this temperature in accordance with saturation conditions—P_s, mass flux density—j and also heat flux density—q.

Thus for determination of four unknown values four equations should be used. First of them is the expression of Fourier thermal conductivity law for one-dimensional case $q = \lambda_{CO_2}\frac{(T_S - T_W)}{\delta}$, where λ_{CO_2} is thermal conductivity of CO_2 deposit. It is considered that the λ_{CO_2} value is known. Second is $q = j \cdot r$, where r is the latent heat of de-sublimation per mass unit. Third is the empirical correlation between saturation pressure and corresponding temperature $P_s = T^{-\frac{1361}{T_S}+12.01}$. Traditional approach suggests in the role of the fourth equation same relationship between saturation pressure and corresponding temperature, that is temperature near the interface is determined in accordance with saturation line. The solution of these four

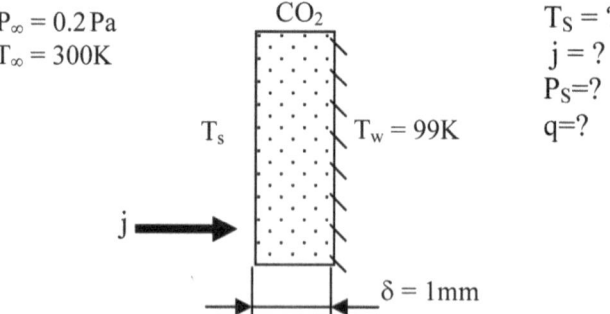

$P_\infty = 0.2\,\text{Pa}$
$T_\infty = 300K$

CO_2

T_s

$T_w = 99K$

$T_S = ?$
$j = ?$
$P_S = ?$
$q = ?$

$j \longrightarrow$

$\delta = 1\text{mm}$

Fig. 2.3 Condensation of gas carbon dioxide (CO_2)

equations gives as a result temperature difference across the deposit layer is about 8 K. But if instead of traditional approach we use the non-equilibrium correlation (2.6) then the value of this temperature difference and mass flux density j becomes in almost fifteen times smaller than at traditional calculations.

References

1. Hertz H (1882) Üer die Verdunstung der Flussigkeiten, inbesondere des Quecksilbers, im lufteeren Raume. Ann Phys Chem 17:177–200
2. Knudsen M (1915) Die Maximale Verdampfungsgeschwindigkeit des Quecksilbers. Ann Phys Chem 47:697–708
3. Risch R (1933) Über die Kondensation von Quecksilber on einer vertikale Wand. Helv Phys Acta 6:128–138
4. Crout PD (1936) An application of kinetic theory to the problems of evaporation and sublimation of monatomic gases. J Math Phys 15:1–54
5. Scharge RW (1953) A theooretical study of inter-phase mass transfer. Columbia University Press, New York
6. Kucherov RY, Rikenglaz LE (1959) On hydrodynamical boundary conditions for evaporation and condensation. J Exp Theor Phys 7(1):125–126 (in Russian)
7. Kucherov RY, Rikenglaz LE (1960) About the measurement of the condensation coefficient. In: Reports (Doklady) of USSR academy of science, vol 133(5). pp 1130–1131 (in Russian)
8. Labuntsov DA (1967) Analysis of the evaporation and condensation processes. Teplofiz Vysok Temper 5:647–653 (in Russian)
9. Muratova TM, Labuntsov DA (1969) Kinetic analysis of the evaporation and condensation processes. Teplofiz Vysok Temper 7(5):959–967 (in Russian)
10. Anisimov SI (1968) Evaporation of metals under the influence of laser radiation. J Exp Theor Phys 54:339–342 (in Russian)
11. Anisimov SI, Imas YA, Romanov GS, Khodyko YV (1970) Effect of high intensity radiation on metals. Nauka Publishing House, Moscow (in Russian)
12. Kogan MN, Makashev NK (1971) On the role of the Knudsen layer in the theory of heterogeneous reactions and flows with the surface reactions. In: Reports of the academy of sciences of USSR: mechanics of liquids and gases, no. 6. pp 3–11 (in Russian)
13. Murakami M, Oshima K (1974) Kinetic approach to the transient evaporation and condensation problem. In: Becker M, Fiebig M (eds) Rarefied gas dynamics. DFVLR Press, Pors-Wahn, paper F6
14. Ytrehus T (1977) Theory and experiments on gas kinetics in evaporation. In: Potter JL (ed) Rarefied gas dynamics. AIAA, New York, Part 2, pp 1197–1212
15. Labuntsov DA, Kryukov AP (1979) Analysis of intensive evaporation and condensation. Int J Heat Mass Transf 22:989–1002
16. Knight SJ (1979) Theoretical modelling of rapid surface vaporization with back pressure. AIAA J 17:519–523
17. Cercignani C (1981) Strong evaporation of a poliatomic gas. In: Fisher SS (ed) Rarefied gas dynamics. AIAA, New York, Part 2, pp 305–310
18. Frezzotti A (1986) Kinetic theory study of the strong evaporation of a binary mixture. In: Boffi V, Cercignani C (eds) Rarefied gas dynamics, vol 2. Teubner, Stuttgart, pp 313–322 (1986)
19. Sone Y, Aoki K, Sugimoto H, Yamada T (1988) Steady evaporation and condensation on a plane condensed phase. Theor Appl Mech 19:89–93 (Bulgaria)

Chapter 3
Evaporation and Condensation of Vapor–Gas Mixtures

It is well known that intensity of transfer processes at evaporation and condensation depends strongly on the presence of a non-condensable gas in vapor–gas mixture. This effect is confirmed by different calculations [1, 2] and experimental researches data [3, 4]. The presence even small gas quantity in the chamber strongly decreases the intensity of evaporation and condensation.

Sometimes the role of non-condensable gas can be very important. For example: in [3] experimental study of the condensation of mercury the small enough pressures were realised. In [4] the concentration of background gas pressure was in five times smaller than that in the previous paper [3]. In these conditions the corresponding mass flux densities at condensation processes differ from each other by 15–20 %.

Usually evaporation–condensation processes with non-condensable component are investigated on the based of traditional approach. In accordance with this approach vapor is removed away from interfaces and supplyed to the surfaces by means of diffusion processes. However, this approach can not be used always. In this case vapor flows can be correctly described on the base of molecular-kinetic approaches which take into account collisions of different molecules. The results demonstrating the importance of accounting for molecule collisions are presented in paper [5].

The application of kinetic methods to the modelling of evaporation–condensation processes have been described in numerous publications. For example, the results of problem solution in case for some sets of component masses and concentrations are presented in paper [6]. In [7] the two-surface problems of a multicomponent mixture of vapour and non-condensable gases in the continuum limit were studied based on asymptotic analysis of the Boltzmann equation. An asymptotic analysis of the linearised Boltzmann equation for the binary mixture was presented in [8]. In paper [9] the direct simulation Monte Carlo (DSMC) method was used for investigated of Couette flow in vapor–gas mixture. The authors of paper [10] studied evaporation and condensation in a vapour-gas mixture on the base of Bhatnagar-Gross-Krook type linearized kinetic equation.

In papers the direct numerical solution of Boltzmann kinetic equation for gas mixture are used for analysed of evaporation–condensation processes. In paper

A. Kryukov et al., *Non-Equilibrium Phenomena Near Vapor–Liquid Interfaces*,
SpringerBriefs in Applied Sciences and Technology,
DOI: 10.1007/978-3-319-00083-1_3, © The Author(s) 2013

[11] technique of joint solution of kinetic equations system and fluid dynamic equations was presented.

Thus the description of evaporation and condensation at the presense of non-condensable gas becomes more difficult in comparison with the case of pure substance. In this chapter at the beginning the evaporation in vapor–gas mixture is studied and then condensation from vapor–gas mixture is considered.

3.1 Liquid Evaporation in Vapor–Gas Mixtures

This problem was investigated by A. P. Kryukov, V. Yu. Levashov, I. N. Shishkova [11, 12].

3.1.1 Problem and Solution Method

Evaporation from interface surface in space occupied by mixture of vapor (water, as example) and non-condensable gas (nitrogen, as example) is considered. Statement of this problem is presented in Fig. 3.1.

Calculative domain is limited on the left side by interface surface. Temperature of this surface is T_1; numerical density of vapor molecules corresponding to this temperature along saturation line is n_{v1}. At the initial time moment all investigated domain is occupied by vapor–gas mixture with numerical densities vapor and gas n_{v0}, n_{g0} correspondingly and temperature $T_0 = T_1$. Velocity distribution functions at the initial moment are prescribed by maxwellians with temperature T_0, corresponding numerical densities and zero mean (macroscopic) velocities.

It is assumed in real physical situation that calculative domain is not limited on the right side as in Fig. 3.1. At numerical investigation there is not possibility to obtain solution namely in semi-infinite space. Therefore, in present study size of calculative domain is taken large enough along x coordinate in order to avoid the disturbance influence of right boundary on the vapor–gas flows near interface surface during the time of solution evolution. In presented cases domain with size about 25,000–50,000 λ_{base} is considered depending on concrete initial and

Fig. 3.1 Statement of a problem

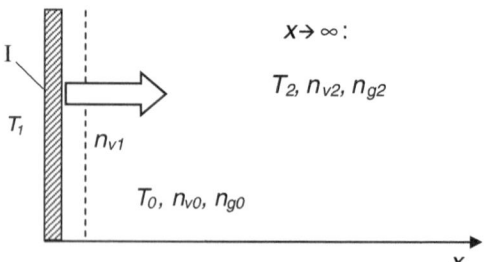

boundary conditions, where λ_{base}—mean free path of water molecules at base parameters of problem.

Base parameters are assumed as temperature $T_{base} = T_1$ and numerical density of water molecules $n_{v\,base}$, corresponding to T_{base} along saturation line. The ratio of component molecular masses is 0.64, the ratio of diameters is 1.25. At $x \rightarrow \infty$ the following mixture parameters are given: temperatures $T_2 \approx T_1$, numerical densities of vapor n_{v2} and gas molecules $n_{g2} = n_{g0}$.

All molecules incident on interface surfaces (limited surfaces) condense on these surfaces. From these surfaces evaporation takes place: n_{v1} и n_{v2}—numerical densities of vapor (water) molecules incoming in calculative domain from limited surfaces. Diffuse type of evaporation is assumed. For gas component both boundaries are impenetrable and reflection is described by diffuse scheme also. All boundary velocity distribution functions are semi- maxwellians with corresponding temperature, numerical densities and zero mean (macroscopic) velocities.

As well known too much computer time is required for the solution of the Boltzmann kinetic equations (BKE) system (3.1) for large calculating range (more then 100 length of the mean free paths for the vapor molecules). As following in this work the technique of joint solution [11] of kinetic equations system and fluid dynamic equations are used for the investigation of evaporation–condensation problem. At this approach system (3.1) is solved in thin domains that adjoin to the surfaces of evaporation (I) (the length of this domain is a few mean free paths of vapor molecules). Mixture flows outside these thin regions are described by fluid dynamic equations (Navier–Stokes) (3.2). In this case the results of the BKE solution are used as boundary conditions for Navier–Stokes equations. This approach gives the possibility to decrease the computer time and to extend the investigated domain.

The BKE system for gas mixtures [13]:

$$\frac{\partial f_v}{\partial t} + \xi_v \frac{\partial f_v}{\partial x} = J_{vv} + J_{vg},$$
$$\frac{\partial f_g}{\partial t} + \xi_g \frac{\partial f_g}{\partial x} = J_{gv} + J_{gg}, \tag{3.1}$$

where $f_v = f_v(x, t, \xi_v)$ and $f_g = f_g(x, t, \xi_g)$—distribution functions for vapor («v») and gas («g») component accordingly, x—coordinate, $\xi = (\xi_x, \xi_y, \xi_z)$—molecular velocity, $J_{\varphi\psi}$ ($\varphi = v, g; \psi = v, g$)—collision integrals.

System of fluid dynamic equations can be presented as following [14]:

$$\frac{\partial \rho}{\partial t} + \frac{\partial \rho u}{\partial x} = 0, \quad \frac{\partial \rho_g}{\partial t} + \frac{\partial \rho_g u}{\partial x} + \frac{\partial j_g^{diff}}{\partial x} = 0$$

$$\frac{\partial \rho u}{\partial t} + \frac{\partial}{\partial x}\left(\rho u^2\right) = -\frac{\partial p}{\partial x} + \frac{4}{3}\frac{\partial}{\partial x}\left(\mu\frac{\partial u}{\partial x}\right) \tag{3.2}$$

$$\frac{\partial \rho e}{\partial t} + \frac{\partial \rho e u}{\partial x} = -p\frac{\partial u}{\partial x} + \frac{4}{3}\mu\left(\frac{\partial u}{\partial x}\right)^2 - \frac{\partial q}{\partial x},$$

where $\rho = \rho_v + \rho_g$—mixture density, $\rho_g = m_g n_g$, $\rho_v = m_v n_v$—densities of «v» and «g» components accordingly, p—mixture pressure, $j_g^{diff} = -\rho D \partial C_g / \partial x$—diffusion mass flux, $C_g = \rho_g / \rho$—mass concentration of «g» component, D—diffusion coefficient, $q = -\lambda_T \frac{\partial T}{\partial x} + (h_g - h_v) j_g^{diff}$—energy flux in binary mixture, h_v and h_g—enthalpy of component, λ_T—thermal conductivity of mixture, e—internal mixture energy, μ—mixture viscosity, m_v and m_g—molecular masses of «v» and«g» component accordingly.

The fluid is assumed to be ideal gas. The model of hard elastic spheres was used for calculation of viscosity, thermal conductivity and diffusion coefficients.

3.1.2 Results and Analysis

Solutions results of the evaporation–condensation problem in semi-infinite space (in the sense described above) for the different contents of non-condensable component and various initial concentration of vapor are shown in Figs. 3.2, 3.3, 3.4, 3.5. All solutions were obtained for $T_0 = T_1$. Results of the evaporation–condensation problem are shown in these Figures in dimensionless form as dependences on coordinate x. At this: $n'_v = n_v / n_{vbase}$, $n'_g = n_g / n_{vbase}$, $x' = x / \lambda_{base}$, $j'_v = j_v / (mn_{vbase}(RT_{base})^{1/2})$. Further everywhere primes are omitted.

Submitted in these Figures distributions are received as result of the non-steady problem solution. Lines presented in these Figures are depicted for time moment when changing density and mass flux density for vapor and gas in the region near interface surface ceased. At this length of such region in accordance with problem solutions is about four hundred mean free paths. It is seen from Figures that near the inter-phase surface thin layer of 5–10 λ in size is formed, in which appreciable change of macroparameters is observed. Outside this layer, i.e. at $x > 10\lambda$, macroparameters are constant right up large enough distance from interface surface (about four hundred mean free paths, as example).

It is interesting to note, that such steady solutions essentially depend on not only from presence in domain non-condensable component, but also on the vapor quantity contained in this area at the initial moment of time. So if to compare results of Figs. 3.2 and 3.3, it is possible to see, that in both cases gas is pushed out by vapor on distance about hundreds mean free path of water molecules λ_{base} at $T_{base} = T_1$ and numerical density of water molecules n_{vbase}, corresponding to T_{base} in accordance with the saturation line. However, the "stationary" (or more correctly "quasi- stationary") vapor density for a case $n_{v2} = 0.01$ makes 0.82, while for a case when initial vapor density is more in 6 times, i.e. $n_{v2} = 0.06$ the "stationary" vapor density is 0.85. Vapor mass flux densities in both cases are close enough and make 0.21 and 0.19 accordingly.

It was noted above that solutions presented in Figs. 3.2, 3.3 were obtained for evaporation–condensation problem at the presence of non-condensable gas. At this

Fig. 3.2 Solution results for
$n_{v1} = 1.1$, $n_{v2} = 0.01$,
$n_{g0} = 0.5$; **a** vapor and gas
density, **b** vapor density,
c vapor mass flux density

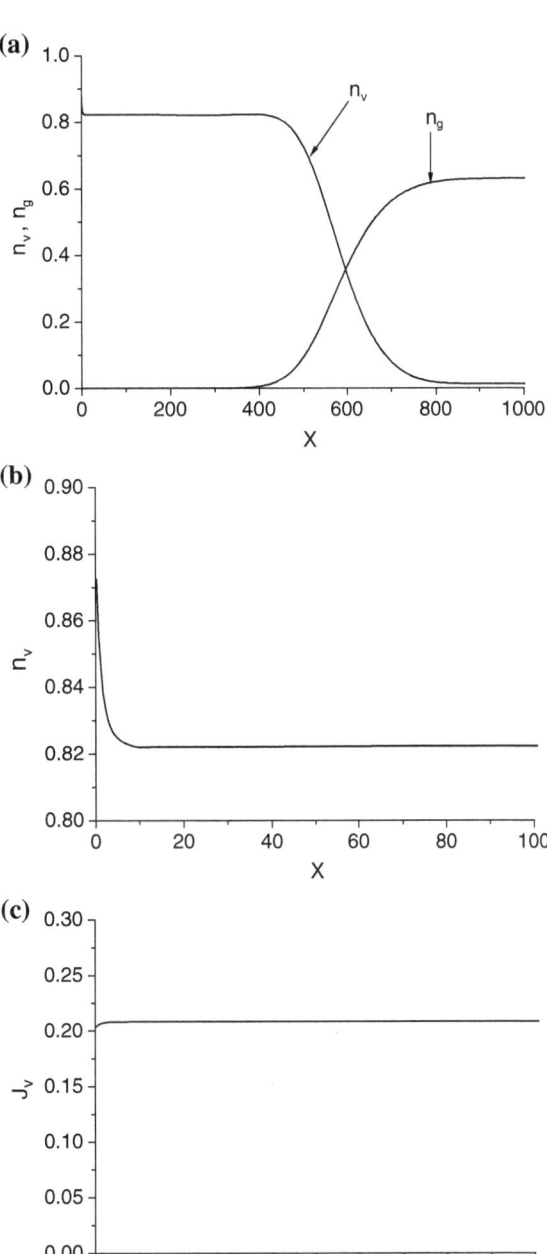

distance between evaporation and condensation inter-phase surfaces was about
25,000–50,000 mean free paths of water molecules λ_{base}. Solutions results for
analogous problem for clean vapor (without gas component) are presented in

Fig. 3.3 Solution results for
$n_{v1} = 1.1$, $n_{v2} = 0.06$,
$n_{g0} = 0.5$; **a** vapor and gas
density, **b** vapor density,
c vapor mass flux density

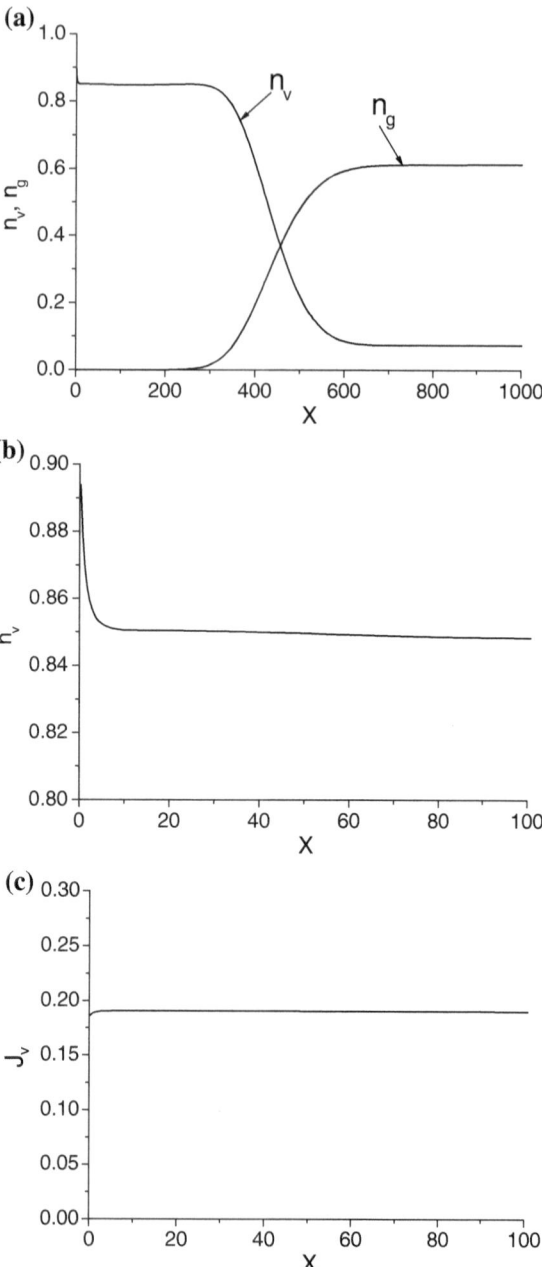

Fig. 3.4. Results in Figs. 3.2, 3.3 seem as similar with data for evaporation–condensation problem in clean vapor qualitatively (Fig. 3.4). However, comparison of Figs. 3.2, 3.3 and Fig. 3.4 shows that the values of vapor and mass flux

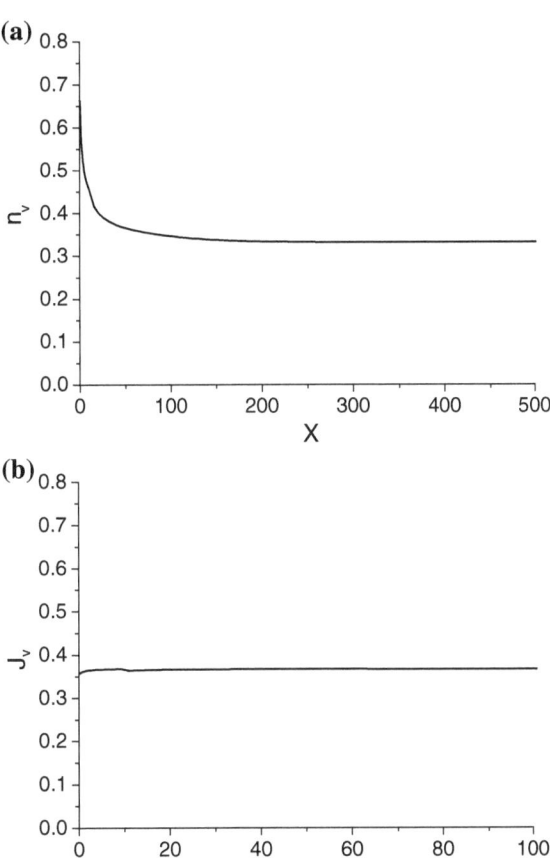

Fig. 3.4 Solution results for $n_{v1} = 1.1$, $n_{v2} = 0.06$, $n_{g0} = 0.0$; **a** vapor density, **b** vapor mass flux density

densities in mixture problems differ substantially from corresponding values for clean vapor.

The discussed solutions are obtained as in and as out the thin limited layer. One can see in Fig. 3.2a or 3.3a that despite of existence of the steady behavior of macroparameters in the region adjoining to surface I, on large distance from it always it is possible to observe front of the solution width 300–500 λ_{base} where vapor and gas densities vary from one constant values up to others. Thus the solution actually represents a step extending along X coordinate. At the same time, there are such ratio of the vapor and gas densities when the solution is qualitatively other. The densities (vapor and gas) are changed from the meanings close to given on interface surface up to the appropriate values at very far from it (Fig. 3.5.). It occurs on the length of the order of several hundreds mean free paths λ_{base}.

If in first two cases (see Figs. 3.2, 3.3) the correlation $n_{v1} \gg n_{v2}$ was valid, further was taken $n_{v1} \approx n_{v2}$ at the same value of initial density of gas n_{g0}: $n_{v1} = 1.1$, $n_{v1} = 1.0$, $n_{g0} = 0.5$ (Fig. 3.5). The solutions presented in Fig. 3.5 differ from that are shown on Figs. 3.2 and 3.3. Change of the vapor density occurs

Fig. 3.5 Solution results for
$n_{v1} = 1.1$, $n_{v2} = 1.0$,
$n_{g0} = 0.5$. **a** vapor density,
b gas density, **c** vapor mass
flux density

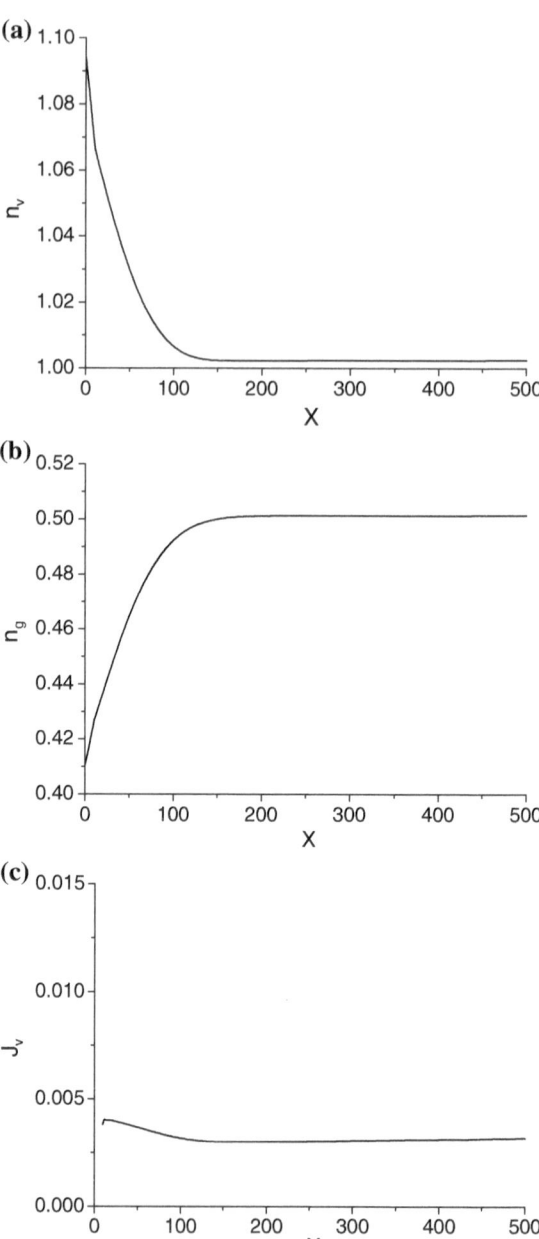

in layer thickness almost 150 λ_{base} instead of 5–10 λ_{base} in the previous problems
(Fig. 3.5a). Complete pushing away from interface surface of the gas it is not
observed (Fig. 3.5b) because intensity of evaporation is too small. The vapor mass
flux density decreases almost in 70 times (Fig. 3.5c). These results shows that

presence of non-condensable gas and initial quantity vapor in investigated domain make essential influence on character of solution.

In order to study the influence of initial gas density on solution behavior the problem was solved at $n_{v1} = 1.1$, $n_{v2} = 0.06$ and different n_{g0}, namely in addition to $n_{g0} = 0.5$ (see above and Fig. 3.3) at $n_{g0} = 0.3$ and $n_{g0} = 0.1$.

Results are presented in Table 3.1. The following designations are used: n_v^* – vapor density according the "shelf" (straight step) in steady solution. In Table 3.1 values n_v^* and vapor mass flux density j_v are given depending on n_{v1}, n_{v2} and n_{g0}.

From this table, first of all, the role of a non-condensable component in researched domain is well visible. The comparison of the solutions No. 1 (Fig. 3.2) and No. 4 shows the following: if amount of the gas near the interface surface (evaporation surface I) equal zero then only vapor borders with interface surface. However, the value of steady vapor density for case No. 1 (evaporation in vapor–gas mixture) in 2.5 times more than similar value for case No. 4 (evaporation in pure vapor). Mass flux density of vapor for case of evaporation in pure vapor (No. 4) in 1.7 times more than for case of evaporation in vapor–gas mixture (No. 1).

Comparison vapor and gas densities dependences on coordinate in Fig. 3.6 and analysis of lines No. 13 and No. 14 in the table shows that approximately at $n_{g0} \approx 0.05$ there is transition between two type of solution: with an without "shelf" near evaporation interface surface for $n_{v1} = 1.1$ and $n_{v2} = 1.0$.

Solution results were obtained for the stage when disturbance from condensation interface surface did not appear in calculative domain. Two solution kinds were revealed. In the first of them gas completely was pushed up by vapor from the region near the evaporation interface surface. At this "shelf" (straight step) in

Table 3.1 Results

No	n_{g0}	n_{v1}	n_{v2}	n_v^*	j_v
1	0.5	1.1	0.01	0.82	0.21
2	0.3	1.1	0.01	0.68	0.29
3	0.1	1.1	0.01	0.46	0.36
4	0	1.1	0.01	0.33	0.37
5	0.5	1.1	0.06	0.85	0.19
6	0.3	1.1	0.06	0.72	0.27
7	0.1	1.1	0.06	0.52	0.35
8	0	1.1	0.06	0.33	0.36
9	0	1.1	0.1	0.41	0.36
10	0	1.1	0.3	0.64	0.30
11	0.1	1.1	0.5	0.24	0.19
12	0.5	1.1	1.0	1.002	0.002
13	0.1	1.1	1.0	1.01	0.009
14	0.05	1.1	1.0	1.02	0.017
15	0	1.1	1.0	1.03	0.036
16	0.5	2	1.0	1.78	0.18
17	0	2	1.0	1.50	0.37

Fig. 3.6 Solution results for $n_{v1} = 1.1$, $n_{v2} = 1.0$, **a** vapor density, **b** gas density

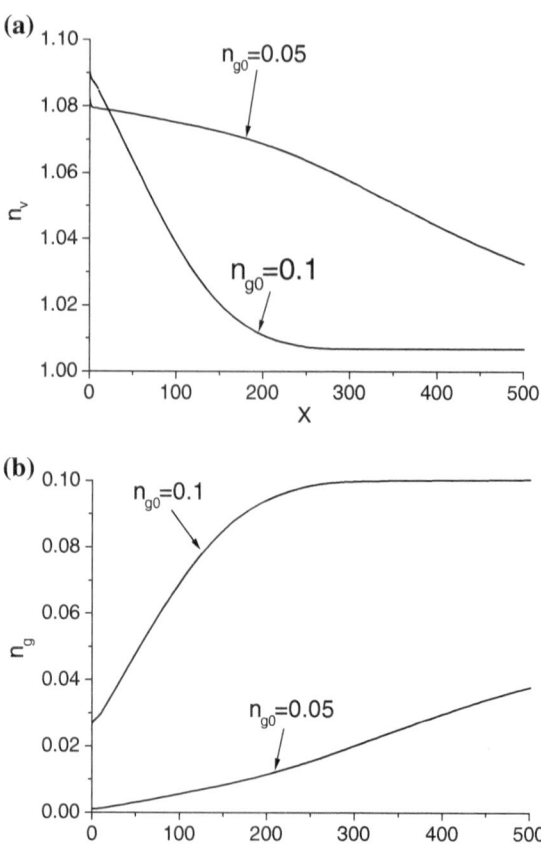

dependence of vapor density on coordinate took place. In the second type such "shelf" was not formatted. Transition from one solution kind to another at prescribed evaporation and condensation interface surfaces temperatures and corresponding vapor saturation densities is determined by the value of initial gas density. For concrete parameters this value (initial gas density at which transition between two types of solutions has place) was found approximately.

3.2 Condensation from Vapor–Gas Mixture

This problem was studied by A.P. Kryukov, V.Yu. Levashov, N.V. Pavlyukevich [2, 15].

3.2.1 Statement of the Problem

Statement of the condensation problem is presented on Fig. 3.7. In this case different statements are considered: condensation of pure vapor (Fig. 3.7a) and condensation at the presence of non-condensable gas (Fig. 3.7b).

As note above usually for this processes investigation the traditional approach is used. This approach is based on the fluid dynamic equations application. In accordance with traditional approach vapor is suppled to interfaces surfaces by means of diffusion. In this case the expression for evaporation–condensation intensity—j—in vapor–gas mixture is obtained on the base of mass conservation equations of each mixture component and impermeability condition for gas $\left(\rho D \frac{dC_g}{dx} + \rho C_g u_x = 0\right)$:

$$j_n = \frac{\rho D}{L} \ln\left(\frac{1 - C_a(x = L)}{1 - C_a(x = 0)}\right). \tag{3.3}$$

where D—diffusion coefficient, L—investigation domain length, C_a—relative densities of evaporation (condensation) component near interface surface and at the distance L from surface.

Nevertheless, it is necessity to note that diffusion model is valid only for case of small vapor or gas concentration. The accuracy of expression (3.3) for the case $\rho_v \approx \rho_g$ is confirmed in paper [16].

However, expression (3.3) can not be used in different real situation success-fully. It is obviously that non-condensable gas can be pushing from interface surface during evaporation process. For example, in technical devices the fol-lowing situation can be realized. In the investigation domain very small quantity of non-condensable component is presented. In this case vapor evaporated from interface surface removes background gas completely. In this case relative density of vapor tends to unity ($C_a(0) \to 1$) and expression (3.3) becomes invalid.

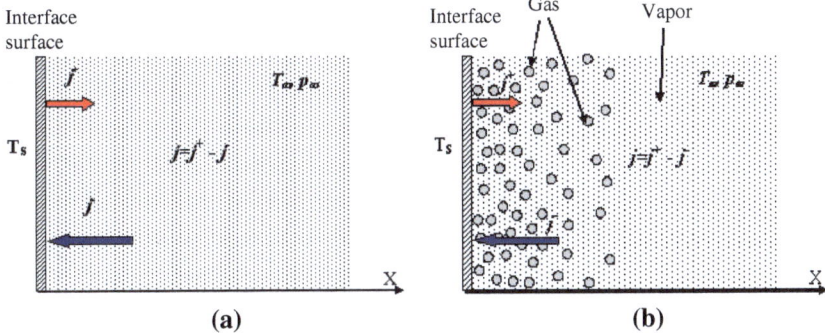

Fig. 3.7 Condensation of pure vapor (**a**), condensation at the presence of non-condensable gas (**b**)

Contrary situation can be realized when non-condensable gas pressure is strongly larger than vapor pressure. These situations take place in vacuum systems for web coating or plasma operations such as sputter deposition, etching, etc. Such vacuum systems typically operate at pressures corresponding to transitional or viscous flow regimes for non-condensable gas, and water vapor cryopumping at free-molecular conditions is highly limited by diffusion of water vapor molecules through a non-condensable gas (argon, air). Traditional approach can not predict this case also [5]. In this chapter the process of water vapor cryopumping through argon is investigated. In this water vapor content is about three orders of magnitude lower than that of the noncondensable gas—argon. The argon pressures range is $0.5 \cdot 10^{-3}$–$20 \cdot 10^{-3}$ torr.

The statement of the problem is following. Two simplified geometries of outgassing and cryocondensing surfaces were taken into account (Fig. 3.8): two infinite parallel flat surfaces, and two infinite coaxial cylinders (outer cylinder is hot outgassing surface, inner cylinder is cold cryocondensing surface). It was assumed that outgassing of water vapor occurs at constant rates in investigation domain.

As note above, usually for this processes investigation the traditional approach is used. However, in this case this approach can not be valid. For example, if argon pressure is equal $P_{Ar} = 2 \cdot 10^{-3}$ torr and water vapor pressure is equal $P_{H_2O} = 10^{-6}$ torr then mixture density is approximately equal to argon density (i.e. $\rho \approx \rho_{Ar}$). In this case relative density is substantially smaller then unity (i.e. $C_{H_2O} = \rho_{H_2O}/\rho \approx 0.22 \cdot 10^{-3} << 1$) and expression (3.3) can be transform to expression (3.3a):

$$j_n = \frac{D}{L} \rho (C_{H_2O}(0) - C_{H_2O}(L)) \tag{3.3a}$$

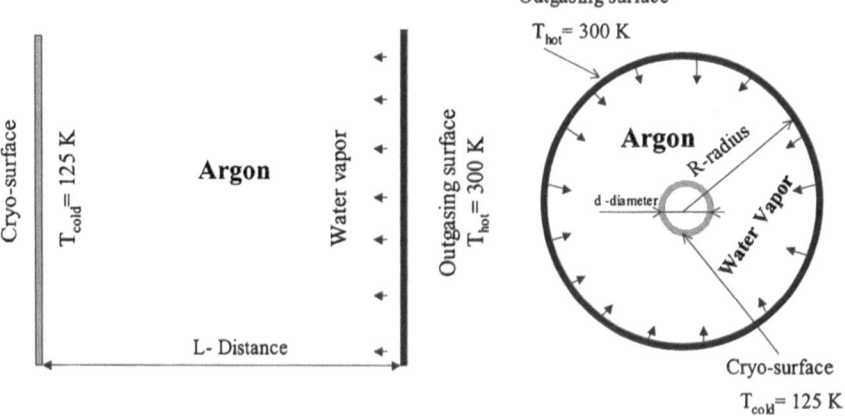

Fig. 3.8 Model

The value of relative density difference can be determined from this correlation in the following way:

$$(C_{H_2O}(0) - C_{H_2O}(L)) = \frac{j_n \cdot L}{D\rho} \qquad (3.3b)$$

As known from experimental measurements the value of vapor flux density is equal approximately 10^{-8} kg/(m^2·s) and if domain size is equal 36 inch (0.914 m), diffusion coefficient for argon-water mixture is $D \approx 3 \cdot 10^{-5}$ m^2/s then $(C_{H_2O}(0) - C_{H_2O}(L)) = 7.135$. Thus this model (diffusion model) cannot predict experimental results because value of relative density can not be more the unit. As shown in paper [5] this flows can be correctly described only by the molecular kinetic theory based on the kinetic Boltzmann equation. The presented below results demonstrate that thorough calculation of molecular collisions allows obtaining of nontrivial results.

3.2.2 Results and Discussion

It is well known that intensity of transfer processes at evaporation–condensation depends strongly on the presence of a non-condensable gas in vapor–gas mixture. This effect is confirmed by different calculations [1, 2] and experimental researches data [3, 4]. For example in paper [2] the condensation process on interface surface is studied for one-dimensional statement when vapor flowed from semi-infinite space throw the binary vapor–gas mixture. In this work the conclusion was made that condensation is possible in principle only if quantity of gas in system is smaller some limited value. This conclusion was made on the base of kinetic approach [17] and with used impermeability condition for gas:

$$\rho D \frac{dC_g}{dx} + \rho C_g u_x = 0$$

In this case the quantity of gas at which condensation process takes place can be found from the following expression:

$$\int_{-\infty}^{0} \rho_g M_\infty d\left(\frac{x}{\lambda_\infty}\right) \leq \rho_\infty \qquad (3.4)$$

where ρ_g is the gas density, ρ_∞ is the mixture density away from condensation surface, M_∞ is the Mach number and λ_∞ is the mean free path of vapor molecules. Thus non-condensable component can lock up the interface surface and condensation stops completely. Similar behavior was noted early by K. Aoki and co-authors in paper [1] at study of vapors flows caused by evaporation and condensation on two parallel plane surfaces at the presence of a non-condensable gas.

It is obviously that the expression (3.4) does not take into account collisions peculiarities between gas and vapor molecules in the near interface surface. In paper [18] this collisions peculiarities are taken into account on the base of model kinetic equation. In paper [18] quantities of gas, at which the condensation process is locked completely, are presented for different Mach number. For example at $M_\infty = 0.1$ condensation ceases if gas quantity is in accordance with following correlation: $\rho_g \geq 9.0\rho_\infty$.

The gas quantity at which condensation process is locked can be found also from

(3.4). In this $\rho_g \approx \rho_g^{\lim} \approx \int_{-\infty}^{0} \rho_g M_\infty d\left(\frac{x}{\lambda_\infty}\right) \approx \rho_\infty$ and if Mach number is equal 0.1

then $\rho_g^{\lim} \approx \int_{-\infty}^{0} \rho_g d\left(\frac{x}{\lambda_\infty}\right) \approx 10.0\rho_\infty$. Thus kinetic calculation and simplified version

(3.4) give similar results.

References

1. Aoki K, Takata S, Kosuge S (1998) Vapor flows caused by evaporation and condensation on two parallel plane surfaces: effect of the presence of a noncondensable gas. Phys Fluids 10(6):1519–1533
2. Kryukov AP, Levashov VY (2008) Condensation from a vapor-gas mixture on a plane surface. High Temp 46(46):700–704
3. Niknejad J, Rose JW (1981) Interphase matter transfer: an experimental study of condensation of mercury. Proc R Soc Lond A 378:305–327
4. Kosasie AK, Rose JW (1992) New measurements for condensation of mercury—implications for interphase mass transfer. In: Proceedings of 7-th ASME national heat transfer conference, San Diego
5. Kryukov AP, Podcherniaev O, Hall PH, Plumley DJ, Levashov VY, Shishkova IN (2006) Selective water vapor cryopumping through argon. J Vac Sci Technol A Vac Surf Films 24(4):1592–1596
6. Pong L, Moses G (1986) Vapor condensation in the presence of noncondensable gas. Phys Fluids 29(6):1796–1804
7. Takata S (1999) Two-surface problems of a multicomponent mixture of vapors and noncondensable gas. Phys Fluids 11:2743–2756
8. Aoki K, Bardos C, Takata S (2003) Knudsen layer for gas mixtures. J Stat Phys 112:629–655
9. Yoshida H, Aoki K (2007) Numerical analysis of the cylindrical couette flow of a vapor-gas mixture in vapour-gas mixtures. In: Ivanov MS, Rebrov AK (eds) Rarefied gas dynamics. Publishing House of the Siberian Branch of the Russian Academy of Sciences, Novosibirsk
10. Onishi Y (1986) The spherical-droplet problem of evaporation and condensation in a vapour-gas mixture. J Fluid Mech 163:171–194
11. Kryukov AP, Levashov VY, Shishkova IN (2007) Evaporation-condensation problem in vapour-gas mixtures. In: Ivanov MS, Rebrov AK (eds) Rarefied gas dynamics. Publishing House of the Siberian Branch of the Russian Academy of Sciences, Novosibirsk. pp 1176–1181
12. Kryukov AP, Levashov VY, Shishkova IN (2009) Evaporation in mixture of vapor and gas. Int J Heat Mass Transf 52:5585–5590
13. Kogan MN (1969) Rarefied gas dynamics. Plenum, New York

14. Labuntsov DA, Yagov VV (2000) Mechanics of two-phase systems. MEI publishing, Moscow (in Russian)
15. Kryukov AP, Levashov VY, Pavlyukevich NV (2010) Condensation from vapor gas mixture. J Eng Phys Thermophys 83(4):637–644
16. Kulikovsky VK, Pavlukevich NV, Vasiliev LL (2008) Mass transfer between evaporator and condenser in drying cell. In: Proceedindg of 3-th international science practical conference "modern energy-saving heat technology-2008", vol 2. Moscow. pp 36–42
17. Labuntsov DA, Kryukov AP (1979) Analysis of intensive evaporation and condensation. Int J Heat Mass Transf 22:989–1002
18. Taguchi S, Aoki K, Takata S (2003) Vapor flows condensing at incidence onto a plane condensed phase in the presence of noncondensable gas. I. Subsonic condensation. Phys Fluids 15(3):689–705

Chapter 4
Motion of Vapor–Liquid Interfaces

There are many researches that concentrated on the evaporation and dynamics of liquid—vapour interface during boiling. Among them as experimental results and as theoretical model with numerical simulation are presented. For example the transient film boiling in the vicinity of a stagnation point on the frontal surface of a very hot blunt body which moves with a constant velocity in an incompressible viscous fluid in the presence of a vapour layer near the body surface is studied in [1]. Within the unsteady two-phase boundary layer approximation, the equations of motion of the liquid and vapour phases are formulated with taking into account the conservation of mass, momentum, and energy on the a priori unknown phase interface. In the vicinity of the stagnation point on the body surface, the solution of the boundary layer equations is sought. At this a parabolic system of partial differential equations is obtained, which is solved numerically. The similarity parameters controlling the film boiling process are determined. On the basis of parametric numerical calculations, the dynamics of the vapour layer are investigated for the case of a plane hot body moving in water with the room pressure and temperature. In the space of governing parameters, the limits of the existence of steady and unsteady film boiling regimes are found.

Some authors presented analytical model of film boiling heat transfer from a sphere in saturated water, but non-equilibrium effects on the vapor—liquid interface are not considered. By analyzing different experiments, it was found that there are in fact two different forced convection film boiling sub-regimes characterized by relatively "low" or "high" heat transfers, and that the existence of these sub-regimes is probably linked with the stability of the vapor film during film boiling [2].

4.1 Motion of Helium II Bridges in Capillary Channels with Vapor at the Presence of Longitudinal Heat Flux

This problem was analyzed by Kryukov and Korolev for ordinary liquids [3, 4], Kryukov and Korolev for helium II [5].

A. Kryukov et al., *Non-Equilibrium Phenomena Near Vapor–Liquid Interfaces*, SpringerBriefs in Applied Sciences and Technology, DOI: 10.1007/978-3-319-00083-1_4, © The Author(s) 2013

In various two-phase systems cases are possible during heat supply when an evaporating liquid flows in a channel filled by vapor. At certain conditions, non-equilibrium effects characteristic of the evaporation and condensation problems may play an important part in the case of such flow. It is the objective of this study to develop a method of predicting the velocity of phase boundaries with regard for the non-equilibrium in vapor in proximity to the interfaces. In accurate common statement the steady flow is not achieved because the bridge length decreases. But at small evaporation intensity quasi-steady motion is possible for which in the first approximation same description is valid as for steady flow.

4.1.1 Statement of the Problem

A cylindrical capillary of diameter d (see Fig. 4.1) is considered, in which a volume of helium II of known length l is located. A heat flux of density q is supplied continuously to the left-hand interface. The side surface of the channel is adiabatic. In the immediate vicinity (from the macroscopic viewpoint) of the right-hand phase boundary, the vapor pressure P_b'' is prescribed, that is P_b'' is known. It is necessary to determine the velocities of He-II normal motion and vapor–liquid interfaces.

4.1.2 Description of the Model, Basic Equations and Results

In accordance with two-fluids model heat transfer in He-II is realized by normal motion. At this temperature difference along helium II much smaller than in ordinary liquid. Therefore problem solution for He-II can differ from results obtained for ordinary liquids principally.

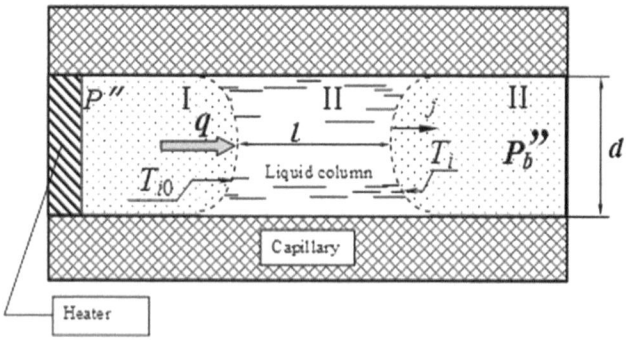

Fig. 4.1 Model

The general conditions of compatibility are formulated, which represent the expressions of the laws of conservation of mass, momentum, and energy on the interfaces [6, 7], as well as special conditions of compatibility describing the peculiarities of transfer processes on the phase boundaries [6, 7]. The special conditions of compatibility are derived using information which is additional to that yielded by the equations of conservation. As a rule, the so—called quasi-equilibrium scheme is employed most frequently, which assumes that the parameters of phases in the vicinity of the interfaces are identical. However, if this equality is strictly required, one can make conclusion about the absence of the transfer processes on the phase boundaries. Therefore, as the intensity of these processes increases, the validity of application of quasi-equilibrium scheme decreases. In this study, another scheme is applied based on solutions of the Boltzmann kinetic equation (BKE) for the problems of heat and mass transfer at evaporation and condensation.

We will treat in succession the respective equations of this description. The expression for the mass flux density j on the right-hand boundary in the coordinate systems related to this boundary is:

$$j = \rho' \cdot (V' - V_i) \qquad (4.1)$$

where ρ' is the density of He-II, V' is the flow velocity of helium II in the laboratory coordinate system, and V_i is the velocity of movement of the right-hand interface. In writing the equation of mass conservation on the left-hand interface, the following should be taken into account. The solution of the BKE for non-stationary problems of heat and mass transfer, in which heat flux is supplied to one heat transfer surface at the initial moment of time and another surface located at some distance from the first one is the vapor liquid interface [8, 9], demonstrates that a steady state characterized by the absence of mass flux through this boundary is attained during the time of kinetic relaxation which is of the order of time between collisions of vapor molecules. Therefore, on a macroscopic time scale, the condition of equality to zero of the mass flux density through the phase boundary being treated has the following form for such problems:

$$\rho' \cdot (V' - V_{i0}) = 0 \quad \text{or} \quad V' = V_{i0} \qquad (4.2)$$

where V_{i0} is the velocity of the left-hand interface. In this expression, it was assumed that the liquid is incompressible and, therefore, moved with the same velocity in proximity to both the left and right-hand interfaces.

The conservation equation for momentum on the right-hand interface at conditions of proportional motion of liquid, assuming that the vapor is an in-viscid medium and the limiting wetting angle is zero, has the form:

$$P_i' + \frac{4\sigma}{d} = P_b'' + \frac{j^2}{\rho_b''} \qquad (4.3)$$

where P_i' is the pressure of liquid in proximity to the right-hand interface (in the region II), σ is the surface tension on the liquid vapor boundary, ρ_b'' is the density of vapor in proximity to the boundary (in the region III)—vapor density near interface (in region III). For the left-hand interface the conservation equation for momentum has the form:

$$P'_{i0} + \frac{4\sigma}{d} = P''$$ (4.4)

where P'' и P'_{i0} are the pressures of vapor and liquid in proximity to the boundary, respectively.

The conservation energy equation for the right-hand interface ("related" to this surface) has the form:

$$j \cdot \left(h'' + \frac{u''^2}{2} \right) + q'' = E',$$

where q'' is the heat flux density in vapor (in the region III) by the thermal conductivity, that is the density of heat flux from to the right interface, h'' is the vapor enthalpy, u'' is the normal velocity of vapor away (from microscopic viewpoint) from the interface, E' is the energy flux density in liquid. Assuming that vapor velocity is small, heat flux density in vapor q'' is close to zero and neglecting cubic in relation velocity addend $ju''/2$, and also supposing that all heat coming to liquid and passing through it is withdrawn owing to liquid evaporation we obtain:

$$j \cdot h'' = E'.$$

The expression for the energy flux density in superfluid helium (He-II) is written in accordance with two-fluid model [10]:

$$E' = \left(\mu + \frac{V_s^2}{2} \right) \cdot j + \rho' STV_n + \rho \cdot V_n(V_n, V_n - V_s)$$

where V_n, V_s are the velocities of He-II normal and superfluid components correspondingly, ρ_n is the normal component density, μ is the chemical potential of superfluid helium, T is the temperature of He-II. If neglecting of cubic relating velocities members is done, then the following correlation is obtained in the coordinate system connected with he right-hand interface:

$$E' = \mu \cdot j + \rho' ST \tilde{V}_n,$$

where $\tilde{V}_n = V_n - V_i$—the relative velocity of He-II normal component, V_i—is the velocity of the right-hand interface in laboratory coordinate system. Substituting this correlation in equality $j \cdot h'' = E'$ and taking into account that $\mu = h - ST$, where r is the heat of vaporization per mass unit (latent heat of evaporation): $r = h'' - h'$, and mass flux density is determined by formula (4.1), we obtain:

$$jr = \rho' ST(V_n - V')$$ (4.5)

where T is, in our case, the mean value of superfluid helium temperature in capillary. Because interfaces temperatures difference is small the following expression is valid:

$$T \approx (T_{io} + T_i)/2 \approx \sqrt{T_{io} \cdot T_i}.$$

Analogy manner is used for deduction of energy conservation equation for the left-hand interface [at this the following equalities: $q'' = q$ and $V_{io} = V'$ from (4.2) in our case are taking into account]:

$$q = \rho' S T (V_n - V') \tag{4.6}$$

Last correlation gives the possibility to find heat flux density in superfluid helium. Taking into account (4.5) and (4.6) the following relationship can be deduced:

$$jr = q \tag{4.7}$$

We write the special condition of compatibility on the right-hand interface as final relations obtained in the solving the steady BKE for evaporation and condensation problems [11, 12]. Accordingly, for linearized formulation, i.e., for:

$$\frac{j}{\rho_S(T_i)\sqrt{2RT_i}} \ll 1, \tag{4.8}$$

these relations have the form [11]:

$$P_S(T_i) - P_b'' = \frac{3}{5} j \cdot \sqrt{2\pi RT_i} \tag{4.9}$$

$$\frac{T_i - T_b''}{T_i} = 0.45 \frac{j}{\rho_S(T_i)\sqrt{2RT_i}} \tag{4.10}$$

$$P_b'' = \rho_b'' R T_b'' \tag{4.11}$$

In (4.9–4.11) R is the individual gas constant of helium, T_b'' is the temperature of vapor in proximity to the interface (in the region III), $P_s(T_i)$ is the saturation pressure corresponding to the temperature T_i. It is necessary to note that correlations (4.9–4.11) are written for the case when coefficient condensation β on the right-hand interface is equal unit. It is easy to make generalization of these relations for the case of arbitrary β values using for example the paper [11]. The special condition of compatibility on the left-hand phase boundary for linearized problem has the form:

$$q = \frac{4}{\sqrt{\pi}} \cdot (P'' - P_S(T_{i0}))\sqrt{2RT_i} \tag{4.12}$$

where $P_s(T_{i0})$ is the saturation vapor pressure corresponding to the temperature T_{i0}.

Note that expression (4.12) is valid for the problems in which one of the heat-transfer surfaces is the condensate vapor interface rather an impenetrable wall. In spite of the fact that, as noted above, in such problems at the stationary stage (on a macroscopic time scale), the mass flux density through the phase boundary is zero, the pressure of vapor differs from that for a similar formulation with two impenetrable for mass flux interfaces. In the latter case, only the heating and expansion of vapor is performed. But in the presence of vapor liquid interface, this surface "follows" the values of density and pressure in the vapor region I owing to evaporation or condensation at the non-stationary (kinetic) stage.

Difference of interface temperatures is connected with pressure difference alone superfluid helium bridge for the case of small deviation from equilibrium by the following thermo-mechanical correlation:

$$\Delta P = \rho' S(T_{i0} - T_i), \tag{4.13}$$

where ρ is the superfluid helium density, S is the entropy of He-II per mass unit. Obviously, that pressure in liquid near left-hand phase boundary is the sum of pressure in liquid near right-hand phase boundary and pressure difference alone superfluid helium bridge:

$$P'_{io} = P'_i + \Delta P = P'_i + \rho' S(T_{i0} - T_i), \tag{4.14}$$

There are eight unknown values:

$$j, P'_i, P'_{io}, T_{i0}, T_i, T''_b, \rho''_b, P''$$

in the set of eight Eqs. (4.3–4.4), (4.7, 4.9–4.12) and (4.14) at prescribed q, d, P''_b and known dependencies $P_s(T_i)$, $P_s(T_{i0})$, $\rho_s(T_i)$. Thus this system of equations is closed and all unknown values can be found.

From this equations correlation for determination pressure difference alone superfluid helium bridge $\Delta P = P'_{io} - P'_i$ is deduced. From Eqs. (4.3) and (4.4) the following relationship (4.15) is obtained:

$$\Delta P = P'' - P''_b - \frac{j^2}{\rho''_b} \tag{4.15}$$

From (4.7, 4.9, 4.12, 4.15), and Clapeyron Clausius equation giving the possibility to find difference of saturation pressures $P_s(T_{i0}) - P_s(T_i)$ at $(T_{i0} - T_i)/T_i \ll 1$ follows:

$$\Delta P = \frac{r \cdot \rho' \cdot \rho''}{T(\rho' - \rho'')} (T_{i0} - T_i) + 0.6j\sqrt{2\pi R T_i} + \frac{\sqrt{\pi} \cdot q}{4\sqrt{2R T_{i0}}} - \frac{j^2}{\rho''_b},$$

where $\rho'' = \rho''(T)$ is the saturation vapor density.

Difference of interface temperatures in this formula is removed by substitution from correlation (4.13) and then ΔP is expressed:

$$\Delta P = \frac{1}{\left(1 - \frac{\rho''}{\rho'-\rho''}\cdot\frac{r}{ST}\right)} \cdot \left(0.6\frac{q}{r}\sqrt{2\pi RT_i} + \frac{q\cdot\sqrt{\pi}}{4\sqrt{2RT_{io}}} - \frac{(q/r)^2}{\rho_b''}\right)$$

$$= \frac{j\cdot\sqrt{2\pi RT_i}}{\left(1 - \frac{\rho''}{\rho'-\rho''}\cdot\frac{r}{ST}\right)} \cdot \left(0.6 + \frac{r}{8RT} - \frac{j}{\rho_b''\sqrt{2\pi RT_i}}\right) \qquad (4.16)$$

$$\approx \frac{j\cdot\sqrt{2\pi RT_i}}{\left(1 - \frac{\rho''}{\rho'-\rho''}\cdot\frac{r}{ST}\right)} \cdot \left(0.6 + \frac{r}{8RT}\right)$$

In order to determine the liquid velocity the equation describing the flow of normal component of superfluid helium is needed. As it has been noted above in common case He-II cannot move uniformly (with constant velocity). However, let us assume the acceleration value at this is small enough to consider that velocity profile is almost same as in Hagen-Poiseuille flow. This motion of liquid can be considered as quasi-steady motion and can be described by steady flow formulae. We will further assume that the laminar mode of normal flow is realized. At these conditions, the known Hagen-Poiseuille relation may be used to describe the flow of He-II normal component [10]:

$$\bar{V}_n = \frac{\Delta P \cdot d^2}{32\eta l} \qquad (4.17)$$

where η is the coefficient of dynamic viscosity of He-II normal motion, $\bar{V}_n$ is the mean in cross section velocity of normal component in coordinate system, connected with capillary walls (laboratory coordinate system).

From Eq. (4.6) it is seen that:

$$V' = \bar{V}_n - \frac{q}{\rho'ST}. \qquad (4.18)$$

ΔP in accordance with correlation (4.16) is substituted in formula (4.17), and this result then is used in relationship (4.18). Thus the expression (4.19) for liquid velocity V' is obtained:

$$V' = \frac{1}{\left(1 - \frac{\rho''}{\rho'-\rho''}\cdot\frac{r}{ST}\right)} \cdot \frac{d^2(q/r)\cdot\sqrt{2\pi RT_i}}{32\eta l} \cdot \left(0.6 + \frac{r}{8RT}\right) - \frac{q}{\rho'ST}. \qquad (4.19)$$

Analysis of this correlation shows that the direction of superfluid helium bridge motion in capillary depends on the length of this bridge. At critical length (let us note this value as L_0) the superfluid helium velocity should be equal zero. Simple reorganization in (4.19) at $V' = 0$ gives the following formula for L_0:

$$L_0 = \frac{\left(0.6 + \frac{r}{8RT}\right)}{\left(1 - \frac{\rho''}{\rho'-\rho''}\cdot\frac{r}{ST}\right)} \cdot \frac{ST}{r} \cdot \frac{d^2\rho'\sqrt{2\pi RT_i}}{32\eta}. \qquad (4.20)$$

It is seen that this value is proportional to capillary diameter in second power and depends on heat flux density very weakly just as dependence on T_i. From (4.19) and (4.20) follows:

$$V' = \frac{q}{\rho' S T} \left(\frac{L_0}{l} - 1 \right).$$

At $l > L_0$ superfluid helium bridge should move to the heater, and at $l < L_0$—from the heater. The proposed model is adequate reality only if between heater and He-II vapor film exists constantly even the thickness of this film is very small (microscopic sizes). On the other side only at the presence of vapor domain superfluid helium will have possibility to move to the heater during long enough time in accordance with above presented description.

Experimental confirmations of this phenomenon in capillary with diameter 250 μm were given in papers [13, 14].

4.2 Evolution of Vapor Films at Boiling of He-II and Ordinary Liquids

This problem was studied by Kryukov A.P., I.M. Dergunov, and A.A. Gorbunov [15].

4.2.1 Statement of the Problem and Basic Equations

This part is devoted to the study of film boiling in superfluid helium. The appearance of film boiling on the heater surface and the recovery to the non-boiling regime have an important significance for fundamental low temperature investigations and cryo-stabilization of various superconducting systems. It is also extremely important for the study of evolution of the system: heater—vapor film—He II in the process of the appearance of film boiling.

Calculations of this process are difficult because the film boiling is strongly non-equilibrium, and thermal resistance is determined mainly by vapor liquid interface processes. Heat transfer is much more intensively realized in He II than in ordinary liquids. Another peculiarity of the problem under investigation is that heat transfer in He II is described by the non-linear equation.

Keeping in mind the mentioned above problems, a combination of continuum medium mechanics and molecular-kinetic approaches are used to consider the problem of the vapor cavity evolution at film boiling of superfluid helium.

A long cylindrical heater of the diameter $d_w = 2R_w$ placed in the He-II bulk at the depth h (Fig. 4.2) is considered. The interface temperature T_b and pressure P_b above it are known. At the moment $\tau = 0$ the load is applied to the heater. The

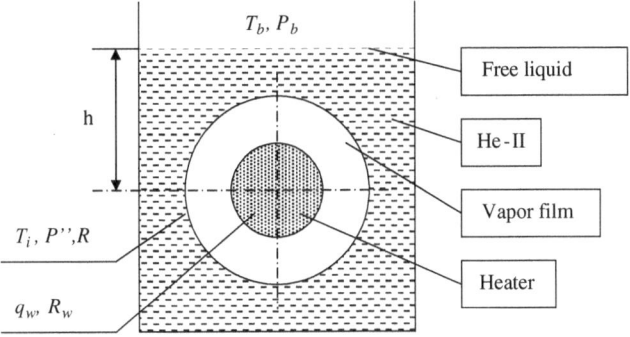

Fig. 4.2 The vapor film on the heater

heat flux density on the heater surface is q_w. The vapor film of the thickness $\delta \to 0$ is supposed to exist from the beginning of the process. The process of the formation of such the film is another important problem which is not considered here. Two opposite scenarios are possible depending on the load supplied.

If the heat load is less than the peak heat flux value, the film collapses and the liquid has a direct contact with the heater. If the heat load is equal to or exceeds the peak value, the vapor film begins to grow. In this case the pressure grows inside the film and the interface temperature increases.

The liquid is considered to be non-compressible. The film thickness is $\delta \ll dw$ during the evolution process. The temperature of the external boundary of the vapor film is usually close to that of a liquid $T_i \approx T_b$, thus the temperature dependence of physical properties can be ignored. The vapor film surface (interphase) is considered to have always a perfect, cylindrical shape.

The boundary motion is described by the ordinary differential equation:

$$\left(R\ddot{R} + \dot{R}^2\right)\ln\frac{R_b}{R} + \left(\dot{R}R\right)^2\left(\frac{1}{R_b^2} - \frac{1}{R^2}\right) = -\frac{\left(P_b + \rho gh - P'' + \sigma/R\right)}{\rho} - 2v\frac{\dot{R}}{R} \quad (4.21)$$

This equation is derived from equations of conservation of mass and momentum for the liquid. In the case of spherical geometry, it is the known Rayleigh equation. It can be presented in the form of two ordinary differential equations of the first order.

Here P_b, is the pressure above the liquid surface p'' is the vapor pressure inside the film, h is the depth of the heater immersion into He-II, σ is the surface tension coefficient, ρ is the liquid density, R is the current vapor cavity radius, $\ddot{R} = \frac{d^2R}{d\tau^2}$, $\dot{R} = \frac{dR}{d\tau} = U_R$ are the acceleration and velocity of the interphase boundary motion, respectively, R_b, is the cryostat inner radius, g is the gravity acceleration, v is the kinematic viscosity of the liquid, for superfluid helium this quantity can be assumed to be an artificial value, in the first approximation v could be equal η_n/ρ, η_n is the viscosity of the normal He-II component. Further the force of mutual friction should be taken into account.

Fig. 4.3 Film boiling on hot body

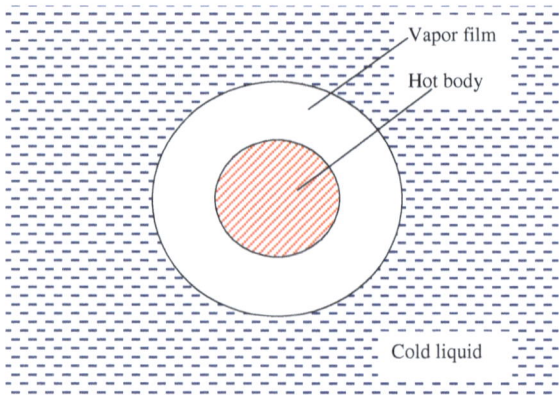

At this place it is necessity to discuss common problem about the system: hot body, vapor film, cold liquid. Schematically this system is presented by the following Fig. 4.3.

If very hot body (cylinder, sphere, heater) is immersed in cold liquid then vapor film is formed around of this body because the film boiling process is realized. It is very interesting to know what process takes place on the liquid vapor interface—evaporation or condensation? It seems on the one side that evaporation should be because the heat from hot body comes to interface. But on the other side owing to the heat from the hot body vapor in the film is heated, temperature of this vapor increases and as result conditions for condensation appear. Accurate answer can make the application of molecular kinetic theory to investigation of this phenomenon.

The vapor phase is described by the Boltzmann kinetic equation (BKE) for linearized problems [11] and later data for non-linear cases [16]. The direct numerical solution of BKE for non-stationary problems [8, 9, 17, 18] has showed that the steady state (stationary stage) in vapor film is reached for the time which is of the order 10, or 100, or 1,000 of the mean free time for vapor molecules. During this time intensive condensation process happens and then mass flux through the interface is ceased (becomes equal to zero). Therefore from the viewpoint of mechanics of continuous media, the steady state including the pressure inside the vapor film is practically immediately established. Thus, the steady, in the molecular-kinetic meaning, correlation between the pressure inside the vapor cavity p'' and heat flux through the vapor liquid interface q at zero mass flux can be used. Such the correlation exists and has the following form:

$$p'' = p_S(T_i) \left[1 + 0.44 \frac{q}{p_S(T_i)\sqrt{2R_\mu T_i}} \right] \tag{4.22}$$

where R_μ is the individual gas constant, p_s is the saturation pressure at the corresponding temperature, q is the heat flux density at the interface.

Equation (4.22) can be applied to analyze the interface heat and mass transfer for both superfluid helium and ordinary liquids. However, for ordinary liquids the heat transfer limitations are established by the convection processes in the liquid bulk. For superfluid helium, the limitation can be determined by processes at the interface. Due to such the behavior, Eq. (4.22) is particularly valid at the analysis of superfluid heat transfer. Equation (4.22) is used for linearized cases.

If $p'' - p_s \sim p_s$, it is necessary to use non-linear correlations (for example [16]) rather than (4.22).

The correct description of non-stationary heat transfer in He-II is a very difficult problem. The authors believe that the best way to solve it is to use hydrodynamics of superfluid turbulence (HST). HST was developed by Nemirovskii et al. [19–21]. The problem of the vapor film evolution should be rigorously investigated by non-equilibrium approaches for both the vapor region (molecular-kinetic theory of gases) and the liquid (HST). Corresponding boundary conditions should be formulated for the interface. In this complicated situation, the emphasis is on the mass momentum and energy transfer at the interface and its motion.

In this connection some simplified models of heat transfer are proposed. The first model is that with the constant liquid temperature, i.e.,:

$$T_i = T_b \tag{4.23}$$

This model, as one can see below, is valid for heaters of small diameters and moderate heat loads.

The second model is that based on the Gorter- Mellink mutual friction theory [22]. Following Seyfert with co-authors and Dresner [23, 24], we assumed that heat transfer in HeII begins immediately in the mutual friction mode observed in the steady state. In other words the heat flux depending on the time and space is determined by the differential Gorter-Mellink law and the classical condition of local energy conservation.

The non-stationary equation of heat transfer in He II for cylindrical geometry has the following form:

$$\rho c \frac{\partial T}{\partial \tau} = \frac{1}{r} \frac{\partial}{\partial r} \left(r \sqrt[3]{\frac{1}{f(T)} \frac{\partial T}{\partial r}} \right) \tag{4.24}$$

where c is the liquid heat capacity, T is the liquid temperature, r is the current coordinate, $f(T)$ is the empirical Gorter-Mellink mutual friction parameter.

The third model uses HST. In the given work we can present the results only for the first and second models.

Initial and boundary conditions are:

$$q = q_w \frac{R_w}{R} \tag{4.25}$$

$$\tau = 0 \quad R = R_w; \quad U_R = 0; \quad q = q_w; \quad T(0, r) = T_b \tag{4.26}$$

$$r = R \quad \left(\frac{\partial T}{\partial r}\right)_{r=R} = -f(T) \cdot q^3; \quad r = R_b \quad T = T_b \qquad (4.27)$$

4.2.2 Results and Discussions

In the first approximation the liquid was considered as a perfect heat conductor. The base for this assumption is the following. Kryukov and Van Sciver [25] have shown that for the steady stage of heat transfer in He-II the vapor liquid temperature for the small heater is closer to the value Tb, than to the value $T_b +$ $\left(\frac{\partial T}{\partial p}\right) \rho g h$. It is related to the fact that the non-equilibrium interphase thermal resistance plays the main role in this case.

The solution of the problem of the evolution with varying the interface temperature is of peculiar interest. Introduction of the non-stationary equation of heat conductivity into the system instead of Eq. (4.24) makes it possible to obtain such the solution for the ordinary liquid. In addition, the solution can be interesting for the study of subcooled He II boiling [26]. Since the vapor film approaches the ordinary liquid (He–I), the evolution problem is also actual for boiling of subcooled He-II. Besides, the non-stationary equation of heat conductivity is simpler than the non-stationary Gorter-Mellink equation and the exact analytical solution with boundary conditions (4.27), i. e., in the problem considered, is known.

4.2.3 The Model with Constant Interphase Temperature

To investigate the geometry influence of the system, the solutions for various inner radius of Dewar helium vessel (cryostat) R_b were obtained in the approximation of perfect heat conductivity (Fig. 4.4). In all the calculations given below $r_w = 38.1 \times 10^{-6}$ m, $q_w = 10^4$ W/m^2 and $T_b = 2$ K, $p_b = 3{,}172$ Pa, $h = 0.01$ m.

As the result of these solutions the time dependencies of R, T_i, p were obtained. As a rule these dependencies occurred to be oscillatory. Upper and lower formats for dependences $R = R(t)$ are presented in Figs. 4.4, 4.5, 4.6.

Film radius oscillations take place between these curves. It is seen that the influence of the cryostat radius on the size of the stabilized vapor film is insignificant. Stabilization time is the only changing parameter. It results from the fact that the larger liquid mass takes a greater time to achieve mechanical equilibrium. On the contrary, an immersion increase results in a significant decrease of the interface oscillation amplitude. At the same time the time of the vapor film stabilization decreases several times.

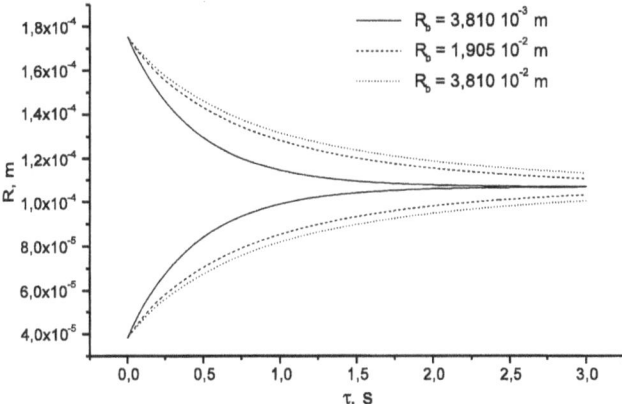

Fig. 4.4 The influence of the system geometry

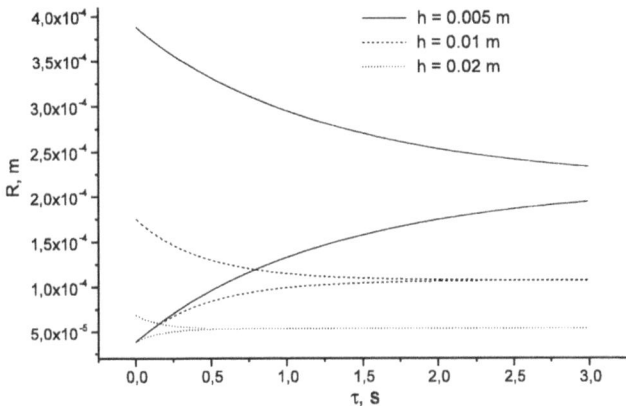

Fig. 4.5 The influence of the immersion depth

Unlike the heater immersion, the increase of the heat load results in the growth of the oscillation amplitude (Fig. 4.6) and the size increase of the film. Stabilization time increases as well.

The dependences presented show that the appearance and growth of the vapor film can be avoided by increasing the external pressure, for example by adding the liquid into the container in terrestrial conditions.

It can be supposed, that these trends remain principally in the models with the variable interface temperature, because the systems of equations differ only by the presence of the equation of heat transfer. It should be expected that time dependences of the vapor cavity radius change due to the effect of the variable interface temperature taken into account.

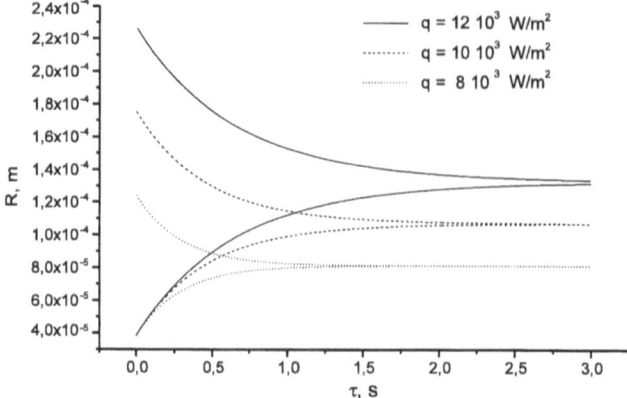

Fig. 4.6 The influence of the heat load value

4.2.4 The Model Accounting for Heat Transfer in Superfluid Helium

Oscillations of the interface temperature were obtained. The maximum values of the temperature corresponds to the minimum values of the vapor film radius and vice versa.

At low heat fluxes occurring at the heater, the influence He II temperature near the interface surface increases insignificantly. The time dependence for R does not change as compared to that obtained using the constant temperature model. Thus the model with the constant interface temperature can be used for small heat loads and it does not lead to a strong distortion of the result. If the heat flux is rather large, the temperature difference $\Delta T = Ti - Tb$ increases significantly. The use of the model with the constant temperature leads to the incorrect results in the case of the large enough heater diameter.

The increase of the interface temperature results in the pressure increase inside the vapor film. The film grows thus to a greater size and at the temperature different from the initial one. The radius of the stabilized vapor film is thus greater than that calculated by the constant temperature model. At the same time it should be noted that due to a very small thermal resistance of He-II, the difference between the models is not significant in a wide range of heat loads at small immersion depths.

4.3 Conclusion

The process of vapor film evolution at superfluid helium boiling has been studied based on the simplified heat transfer models. The approach proposed is based on combination of the molecular-kinetic theory of gases and methods of mechanics of continuous media. The model with the constant liquid temperature makes it possible to obtain the results for the small heater diameter which, in the authors' opinion, are close to the real situation. It can be explained by a very effective heat transfer in He-II. The influence of various parameters (cryostat dimensions, depth of immersion, heat load value) has been investigated. The qualitative relations of the system behavior have been found.

References

1. Vargaftik GM, Osiptsov AN (2011) Film boiling on a hot body moving in a fluid. Fluid Dyn 46(6):942–952
2. Malmazet E, Berthold G (2009) Convection film boiling on horizontal cylinders. Int J Heat Mass Transf 52:4731–4747
3. Kryukov AP (2000) The flow of liquid in a channel with vapor in the presence of longitudinal heat flux. High Temp 38(6):909–913
4. Korolev PV, Kryukov AP (2001) Unsteady-state flow of liquid of high thermal conductivity in a vapor-filled capillary in the presence of longitudinal heat flux. High Temp 39(2):315–321
5. Korolev PV, Kryukov AP (2002) Superfluid helium motion in a capillary with vapor at the presence of longitudinal heat flux. Vestnik MEI 1:43–46 (in Russian)
6. Labuntsov DA, Muratova TM (1972) The physical and methodic principles of formulation of problems of heat and mass transfer under conditions of phase transformations. Teplo- i massoperenos (Heat and Mass Transfer), Minsk: ITMO AN BSSR (Inst. of Heat and Mass Tansfer, Belorussian Acad.Sci.) 2, part 1, p 112
7. Labuntsov DA, Yagov VV (2000) Mechanics of two-phase systems. MEI publishing, Moscow in Russian
8. Kryukov AP, Yastrebov AK (2003) Analysis of transfer processes in vapor film at the interaction of strong hot body with cool liquid. High Temp 41(5):771–778
9. Kryukov AP, Yastrebov AK (2006) Heat and mass transfer through vapor film with taken into account the motion of liquid-vapor interface and rise of interface temperature. High Temp 44(4):556–564
10. Landau LD, Lifshitz EM (1986) Gidrodinamika. Teoreticheskaya fizika (Hydrodynamics: Theoretical Physics), Nauka, (in Russian)
11. Muratova TM, Labuntsov DA (1969) Kinetic analysis of the evaporation and condensation proeesses. Teplofiz Vysok Temper 7(5):959–967 (in Russian)
12. Labuntsov DA, Kryukov AP (1979) Analysis of intensive evaporation and condensation. Int J Heat Mass Transf 22:989–1002
13. Korolev PV, Kryukov AP, Mednikov AF (2005) Experimantal study of He-II motion in capillaries. In: Proceedings of XV school-seminar of young scientists and experts under the direction of the the member of Russian academy of sciences A.I. Leontiev, vol 1. MEI publishing, Moscow, pp 216–219 (in Russian)
14. Korolev PV, Kryukov AP, Mednikov A (2006) Experimental investigation of He-II motion in capillary at the presence of vapor cavity near heater. Vestnik MEI (4):27–33 (in Russian)
15. Dergunov IM, Kryukov AP, Gorbunov AA (2000) The vapor film evolution at superfluid helium boiling in conditions of microgravity. J Low Temp Phys 119(3/4):403–411
16. Khurtin PV, Kryukov AP (2000) Some models of heat transfer at film boling of suoerfluid helium near Λ-point in microgravity. J Low Temp Phys 119(3–4):413–420
17. Kryukov AP, Shishkova IN (1997) Transfer phenomena in vapour film on the interphase of superfluid helium in terrestrial conditions and in microgravity. In: Proceedings of joint Xth

European and VI-th Russian symposium on physical sciences in microgravity, vol 1. Moscow. p 341

18. Kryukov AP, Shishkova IN (1997) Heat-mass transfer through the vapour film at boiling of superfluid helium. Abstracts of Chernogolovka workshop on low temperature microgravity physics (CWS-97), Chernogolovka, p 14

19. Nemirovskii SK, Lebedev VV (1983) The hydrodynamics of superliquid turbulence. Sov Phys JETP 57:1009

20. Nemirovskii SK, Tsoi AN (1989) Transient thermal and hydrodynamic processes in superfluid helium. Cryogenics 29:965

21. Nemirovskii SK, Kondaurova LP, Baltsevich AJ (1994) Unsteady heat transfer in He II with cylindrical geometry. Cryogenics 34:313

22. Gorter CJ, Mellink JH (1949) On the irreversible processes in liquid helium II. Physica 15(3–4):285–304

23. Seyfert P, Lafferranderie J, Claudet G (1982) Time dependent heat transport in subcooled superfluid helium. Criogenics 22:404

24. Dresner L (1984) Transient heat transfer in superfluid helium. Part II. Adv Cryo Eng 29:323–333

25. Kryukov AP, Van Sciver SW (1981) Calculation of the recovery heat flux from film boiling in superfluid helium. Cryogenics 21(9):525–528

26. Murakami M, Yamaguchi M, Yanase N, Ihaba H (1998) Adv Cryo Eng 43:112

Chapter 5
Liquid–Vapor Interface Form Determination

This problem was investigated by Kryukov and Puzina Yu. Yu in [1, 2].

The method for calculation of liquid–vapor interface shape at the nitrogen film boiling on the sphere is presented. This model based on the approach of molecular-kinetic theory for boundary condition formulation. The numerical solution of the equations system has been obtained. Then experimental and calculation data are compared. The results for another liquid were received. The time dependence and mass drop influence are discussed.

This part proposes a calculation method for the axisymmetric form of the vapor–liquid interface at the water drop floating in liquid nitrogen. Investigation of transport processes features in the vapor film is based on the methods of continuum mechanics and molecular-kinetic theory. Non-equilibrium boundary condition is used to describe the effects near the interface. Interface shape is determined by integrating the equation for the pressure difference between vapor and liquid. The heat flux and mass flux due to the evaporation influence on the interface shape.

Using of non-equilibrium boundary conditions gives the possibility to solve difficult problems taking into account peculiarities of the transport processes at the vapor–liquid interface. In practice, in many important cases the interface is axis-symmetric [3]. The solution of the hydrostatics equation with using effect of surface tension and gravity and appropriate boundary conditions gives the equilibrium shape of the interface. However, not all decisions are in fact possible to observe in practice. In addition to hydrostatic equilibrium interface should be stable, at least for small deviations from the equilibrium state shapes. Full analysis includes the parametric studies of the problem, the study of existence range and stability of solutions, as well as the determination of the applicability limits of the model.

An interface shape and processes of heat and mass transfer are interrelated. At the film boiling heat is supplied to the interface from the vapor (through vapor film), and the heater temperature can exceed the critical temperature of the working liquid. Axisymmetric interface at the immersion heater in the liquid is determined by the highly non-equilibrium conditions in contrast to the problems of hydrostatic equilibrium, where heat source does not exist.

A. Kryukov et al., *Non-Equilibrium Phenomena Near Vapor–Liquid Interfaces*, SpringerBriefs in Applied Sciences and Technology, DOI: 10.1007/978-3-319-00083-1_5, © The Author(s) 2013

Formation and growth of vapor bubbles on the heated surface immersed in liquid are studied for a long time. For the calculation of these processes the different models are offered. However, as the rule, in these models it is accepted that the heat is transferred to the interface surface vapor–liquid from the liquid side. In some cases for example at the film boiling heat transfer to the interface surface goes through the vapor film. Thus the heat is brought to the vapor–liquid boundary from vapor part.

Knowledge of the laws of the transition from film boiling regime to bubble (second boiling crisis) is important for many industrial processes [4]. A special role this phenomenon plays at the initiation of vapor explosion caused by spontaneous fragmentation of hot drops immersed in the cold liquid. It is generally accepted that the mechanism of breaking of these drops is associated with the explosive destruction of the vapor film and the instantaneous contact of hot and cold heat transfer [5, 6]. In contemporary literature there are many theoretical and computational works describing phenomena preceding and accompanying the initial stage of contact (see, e.g., [7, 8]). However, details of the physical processes, such as shape and influencing factors on the interface, studied insufficiently.

The formulation of boundary conditions play a main role at the solving of the fluid dynamics equations by the numerical methods at the significant heat and mass transfer through the interface surface. The most complete description of boundary processes can be achieved through the use of molecular-kinetic methods (for example see above Chaps. 2 and 3). A similar approach can be successfully used for mathematical modeling of the phenomena that accompany the process of destruction of the vapor layer near the heated surface and accurate interpretation of the experimental data [9].

5.1 Statement of the Problem and Model

Liquid cryoagent is used as a heat sink at the freezing of liquid drops for high intensity of the cooling process [10]. At this situation when droplets near the cryoagent surface have a higher temperature than the limiting (biggest) of liquid overheating may be realized. This leads to the following physical effects: drops with a density greater than the density of the liquid cryoagent float on its surface long enough time (several tens of seconds), surrounded by a vapor film.

We propose the following physical model of steady floating water droplets in liquid nitrogen. Sphere from water of radius R_w floats in the saturated liquid nitrogen. Pressure over the free surface of liquid P_b is known (Fig. 5.1). Stable vapor film is formed due to the high temperature of the drop T_w. Axisymmetric shape of the vapor–liquid interface is described in a cylindrical coordinate system z, r. The lower (front) point of the interface located on the vertical axis of the sphere is considered as the coordinate origin.

Distance from the origin to the free surface of the liquid h_0 and immersion depth of the drop h_w are determined by the problem solution. Changes in vapor

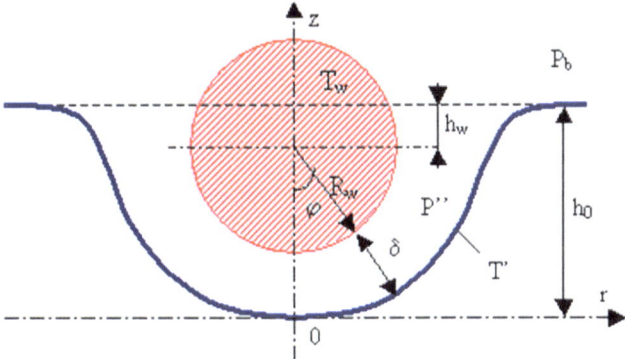

Fig. 5.1 Model

pressure P'' and the thickness of vapor film δ are considered depending on the angle between the vertical axis and the segment from the center of the ball to the current point of vapor film.

As can be seen from the scheme of the problem, the liquid is separated from the heater by the vapor film. The heater is water drop in this case. Heat comes to the interface from the vapor side. Liquid is evaporated from the surface owing to the action of the heat flux. Heat flux and mass flux are realized in opposite directions. Vapor film is connected to the outside volume near the free liquid surface. Interface is nearly spherical shape at the bottom. Vapor moves in the channel formed by a permeable for mass flux vapor liquid cryoagent interface and a heater.

Assumptions of physical model are accepted for formulation of the mathematical description. Heat mass transfer processes in two-phase system are considered as quasisteady. The condensation coefficient on interface surface is equal to unit. Temperature of interface surface T' is constant on cross section. Saturation pressure corresponding to temperature of liquid $P_s(T')$ is equal to pressure over a free surface of liquid P_b. Vapor flow is laminar. The heat flux by radiation is neglected. Physical properties of a liquid and vapor are constant. Temperature jumps on the interface surfaces of vapor–liquid and vapor–heater wall are smaller in comparison with the general difference of temperatures. Possible fluctuations of vapor film are not considered.

The drop floats on the liquid surface. The force of gravity is balanced by the action of vapor pressure difference above and below the object. The balance of forces acting on the ball allows us to determine the depth of his immersion h_w:

$$m_w g = 2\pi R_w^2 \int_0^\pi P''(\varphi) \sin\varphi \, \cos\varphi \, d\varphi \tag{5.1}$$

The mass balance in vapor film is the following:

$$\int_0^\varphi jR_1^2 \sin\varphi d\varphi = \rho'' \overline{w''} R_1 \sin\varphi\delta \tag{5.2}$$

$R_1 = R_w + \delta$—the distance from the heater center to the point on the interface (local film radius). At this vapor film thickness is determined by the following expression:

$$\delta = \sqrt{(h_0 - h_w - z)^2 + r^2} - R_w$$

Equation governing the motion of vapor in the film for laminar flow is the following:

$$\frac{dP''}{ds} = \frac{\xi\eta''\overline{w''}}{\delta^2} \tag{5.3}$$

Heat flux on the vapor–liquid interface is determined by thermal conduction:

$$q_1 = \frac{\lambda''(T_w - T_1)R_w}{\delta R_1} \tag{5.4}$$

From the equation of momentum conservation for liquid the next expression is followed:

$$P' = P_\infty + \rho'g(h_0 - z) \tag{5.5}$$

The Eqs. (5.2–5.5) are added by the laws of conservation of mass, momentum and energy at liquid–vapor interface.

The consideration of interface curvature change leads to the necessary using the conservation law of normal component momentum:

$$P'' - P' = 2\sigma K \tag{5.6}$$

The shape of liquid–vapor interface can be determined from this equation. At this the curvature for axis symmetric case is written in the form of nonlinear differential operator of second order [11].

The boundary conditions for the differential equation solution are expressed in the following form:

$$z = 0, \ r = 0, \ dz/dr = 0 \tag{5.7}$$

At the same time the integration of Eq. (5.6) determines one-parameter family of surfaces curves. One from this family should be chosen as the satisfying to boundary condition on the free liquid surface:

$$z = h_0, \ dz/dr = 0 \tag{5.8}$$

The balance of heat on the surface is following:

$$q_1 = jL \tag{5.9}$$

Non-equilibrium boundary condition for the phase transition in one-component system (according to the linear theory [12]) at the condensation coefficient equal one is written in the following form:

$$\frac{P'' - P_s(T_1')}{P} = -1.2\sqrt{\pi}\frac{j}{\rho''\sqrt{2RT}} + 0.44\frac{q_1}{P''\sqrt{2RT}} \tag{5.10}$$

It is seen from this expression that the actual pressure in vapor is connected with heat flux coming from the heater to the interface and temperature of liquid in accordance with the saturation line.

Thus, the closed mathematic description of the interfacial problem, which purpose is determining of the vapor–liquid interface shape as a dependence $z(r)$, is formulated. It satisfies conservation laws of mass, momentum and energy.

The vapor pressure difference between the bottom point and the free liquid surface is determined on the one hand by the nonequilibrium effects near the liquid–vapor interface on the other hand the hydrostatic pressure of liquid and Laplace "jump" at the interface. The resistance of the viscous friction due to the flowing of vapor in the channel formed by a permeable interface and a heater determines the stability of vapor film. The vapor film is stable if the vapor pressure difference due to nonequilibrium effects near the interface is equal to the pressure difference owing to vapor flow friction of channel formed heater and interface surface. When the heat flux diminishes and hydraulic equilibrium is disturbed, the vapor film collapses, and drop sinks.

Solving of Eqs. (5.1–5.10) gives the possibility to determine not only the shape of the interface, but also the distribution of vapor parameters on the cross section and heat transfer efficiency.

5.2 Comparison with Experimental Data

Heat transport phenomena during floating of hot spherical object in cold liquids were investigated experimentally [10]. In this work cooling of various substances in liquid nitrogen was considered. In particular the water drops were studied with initial temperature in various experiences from 290 to 340 K. Thus, drop temperature was in several times more than temperature of a cooling liquid (nitrogen) ~ 78 K. The corresponding difference became $\Delta T = 212$–262 K, therefore water was hot object for the cryogenic environment.

The water drop with a diameter of 3 mm was floating in liquid nitrogen at atmospheric pressure, the liquid temperature $T' = 77.4$ K. The temperature drops $T_w = 340$ K exceeds the critical temperature of nitrogen $T_{cr} = 126.2$ K. The results of the calculation (Fig. 5.2) show good agreement with data [10]. The lower point of

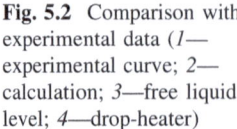

Fig. 5.2 Comparison with experimental data (*1*—experimental curve; *2*—calculation; *3*—free liquid level; *4*—drop-heater)

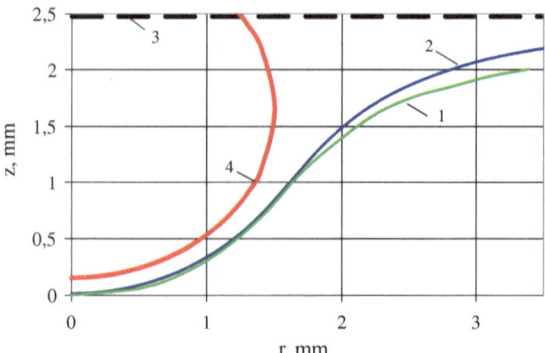

the experimental curve was chosen for the origin. Calculative values of heat flux and heat transfer coefficients to nitrogen are agreed with experimental data well. Thus the developed physical model shows good enough correspondence to the real experimental situations.

The heat and mass transfer processes characteristics for this three-phase system have been calculated. They are presented below. Calculation of the second critical heat flux density for nitrogen gives value 1.36×10^4 W/m^2. Interfacial heat flux is 3.25×10^4 W/m^2 in accordance with thermal conductivity Fourier law for the frontal point at $T_w = 340$ K. Corresponding value of the heat transfer coefficient is 150 W/(m^2·K). The radiation component of heat flux from the heater to the interface is 7.68×10^2 W/m^2, this value is not exceed 5.65 % of heat flux by thermal conductivity. The value of 189 W/(m^2·K) was obtained as a calculation result of the heat transfer efficiency at the film boiling on immersed spheres [4]. The calculations of temperature difference at the vapor–liquid and vapor—the heater interfaces gives the value 1.7×10^{-3} K and 0.36 K respectively with a coefficient of energy accommodation 0.5 in the latter case.

5.3 Results Discussion

The main calculation result is a curve that characterizes the form of the vapor–liquid interface. The further analysis is based on the curves received at different initial data. In addition, the thickness of vapor film and the heat flux from the heater are important values.

The main difference of the problem under consideration from the classical hydrostatic equilibrium problem finding the shape of the interface is connected with influence of the heat flux on the formation of the interface.

The calculation results of the shape of an axisymmetric vapor–liquid interface at the floating of hot ball in cold liquid for the scene selected initial data are presented in Fig. 5.3. A special feature of the solution is that curvature in the lowest point of the interface K_0 is unknown as a parameter of integration.

Fig. 5.3 Shape (*1*) and curvature (*2*) of axysymmetric interface. *3*— free liquid level

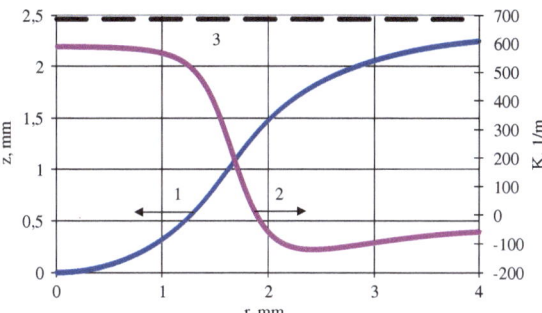

Obviously, the surface shape is spherical in this point, but the radius of curvature R_0 is determined in the solution process. The second boundary condition is given for the liquid free surface, the derivative dz/dr is equal zero. The parameter K_0 is chosen that in addition to the boundary conditions, the distance from the top of the curve $z(r)$ to the free liquid level was minimal. As a result of this selection for the above initial data the radius of curvature of the frontal point was obtained as $R_0 = 1.7$ mm, the distance from the lowest point of the interface to the level of the liquid free surface was $h_0 = 2.46$ mm.

The heater immersion depth h_w is determined as the results of integration to satisfy Eq. (5.1)—the mechanical equilibrium of the drop. Mass of sphere (a drop of water) is estimated as $m_w = 1.4 \times 10^{-5}$ kg, which gives the depth of immersion $h_w = 0.82$ mm.

The shape of liquid–vapor interface has a sophisticated outline with a point of an inflection and a smooth exit on a free liquid surface. The feature of this problem is appearance of the analyzed object as the result of heat flux action unlike another situation discussed below.

The problem concerning the axis symmetric geometry of vapor formation is hydrostatic problem. But usually in such type of problems there are no heat flux sources. At this curvature of vapor–liquid interface K increases as approaching to the surface of solid and the bubble appears inside a sphere with the radius corresponding to curvature in bottom point K_0 (see a dashed line in Fig. 5.4 [3]).

In considered case (Fig. 5.1) vapor–liquid interface curvature K at the removal from the bottom point decreases. Thus, the sphere with the radius corresponding to curvature in bottom point K_0 disposes in a vapor.

Difference between problems with a heat flux source and hydrostatics problems is absence of traditional concept of a contact angle. The interface smoothly leaves on a free level of liquid, thus curvature K changes a sign. As a result of the numerical solution there is a family of the curves describing the shape of liquid–vapor interface. The solution from this family which does not cross level of a liquid free surface and asymptotical exist on this level should be chosen as correct. The distance from a bottom point to a liquid free surface $h_0 = R_w + \delta_0$ also depends on choice K_0 and is determined as a solution result.

Fig. 5.4 Bubble under the
horizontal solid surface

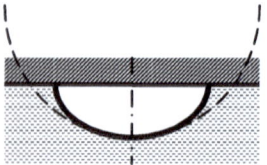

It should be noted that the force acting on the drop from the vapor is an integral characteristic. The calculations for different values have been made to determine the immersion depth of the drop. The overall calculation of this problem is specific algorithm. Minimal changes in initial data lead to corresponding changes in the heat transfer processes characteristic defining the equilibrium of the hot drop floating on the surface of the cold liquid and vapor film surrounding the object. Thus the calculation for range of the parameters is needed.

In order to calculate the unsteady problem of hot object floating in a cold liquid joint solution of the equations system describing the processes of heat and mass transfer at the interface vapor–liquid, and a differential equation of heat balance in the heater should be made. At the same time in each stage it is necessary to compare the temperature of the hot object with an initial boiling point and the minimum value at which the floating can be realized.

5.3.1 Immersion Depth Dependence on Drop Mass

We consider the situation where the mass of the floating object is varied, so another substance than water falls in nitrogen with a greater or lesser density. The dependence of the immersion depth h_w and thickness of vapor film in the frontal point δ_0 on the floating object mass m_w is shown in Fig. 5.5. When the mass increases the depth of immersion increases also, and the thickness of vapor film decreases. This comes from the fact that the pressure under the drop is greater, because heat flux at the interface rises.

Fig. 5.5 Mass drop
influence: *1*—drop
immersion depth; *2*—vapor
film thickness in bottom point

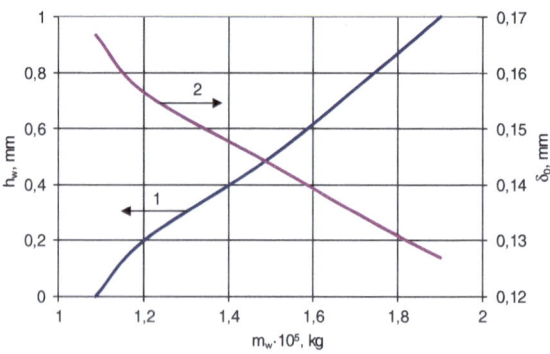

Fig. 5.6 Vapor film
thickness (*2*) and drop
temperature (*1*) dependences
on time

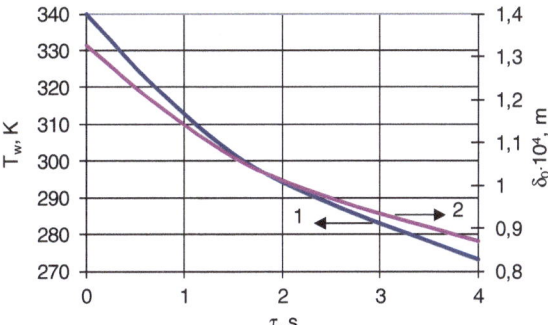

It should be noted that the increase in droplet size also leads to a corresponding transformation of the interface form. Effect of surface tension is diminished, the drop immersion depth increases due to increased mass of the heater, until it reaches a critical value. Then all drops could not be on the surface of the liquid regardless the initial temperature.

5.3.2 Vapor Film Thickness Dependence on Time

Cooling of the drop is calculated from the heat balance. Fig 5.6 shows vapor film thickness (2) dependence on time and corresponding variation of drop temperature (1). Experimental observations give the following. As soon as the heater temperature at the cooling becomes less than a certain size, vapor film on the surface of the hot body collapses and drop sinks. The total supporting force acting on the drop decreases due to the heat flux decreasing at the reducing of the temperature difference between the surface of hot sphere and the liquid. Consequently, the mechanical equilibrium (5.1) of drop is broken, that was observed in the experiments.

The floating time was 20 s for observation with initial data as in Fig. 5.5. The starting time of crystallization was about 5 s. The physical model does not take into account the phase transition inside the drop, the water temperature is limited by the beginning of solidification. Dependence has some deflection due to the changes in drop immersion depth h_w.

5.3.3 Liquid Properties Influence

The quantitative characteristics of the heat mass transfer processes through the vapor–liquid interface depend on thermophysical properties of liquid (Table 5.1). Initial data are chosen in such a way that the thicknesses of vapor film in the frontal point coincided. In this case all other parameters are significantly different.

Table 5.1 Liquid properties influence

Liquid	Nitrogen	Water	Sodium
T_w, K	340	760	2,600
T', K	77.4	373	1151.2
$q_w \cdot 10^{-5}$, W/m^2	3.93	2.34	6.36
$m_w \cdot 10^5$, kg	1.41	4.35	4.99
$\delta_0 \cdot 10^4$, m	1.38	1.37	1.36
$(P''_0 - P_b)$, Pa	30	92.7	124

Fig. 5.7 Liquid–vapor interface variation: *1*—nitrogen; *2*—water; *3*—sodium; *4*—drop (heater); *5*—free liquid level

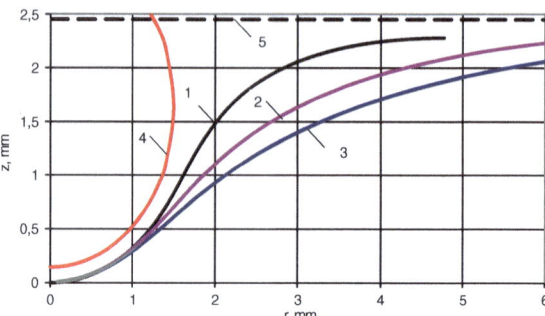

Figure 5.7 shows the dependence of the coordinates describing the geometry of the interface. The thickness of vapor film δ depends on the choice of the integration parameter K_0. For sodium and water the size of the deformed region is several times greater than for nitrogen. This is associated with a greater value of surface tension. At this heat load is less than for nitrogen, whereas the vapor pressure is greater. This feature is due to the nonequilibrium effects on the interface and the influence of evaporation and, consequently, the heat of vaporization of the total pressure difference in a vapor between the lower point and the level of free liquid surface.

Another similar problem is researched at the study of the subcooled liquid boiling. In particular, vapor film is considered on a hemispherical heater immersed in subcooled water [9]. The differences from the previous problem are the following. First, the liquid nitrogen was at saturation condition but not the subcooled water. The second one is that hot object lies on the free surface level and the equation of mass balance was used consequently. The main features of heat mass transfer processes at the problem of film boiling on a hemispherical heater in subcooled water are described in [2].

5.4 Conclusion

The calculation method for the shape of the stationary axisymmetric vapor–liquid interface formed during immersion heater ball into the liquid was presented. The description of the transport processes at the vapor–liquid interface was based on the system of equations including nonequilibrium boundary condition. As a result of the numerical solution the dependence of the interface deflection on the distance from the axis of the heater was obtained for the water droplets with a diameter of 3 mm at room temperature, floating in liquid nitrogen at atmospheric pressure. It was shown that the curvature of the permeable for the mass flux interface depends on the heat flux. Comparison of calculation results with experimental data shows satisfactory agreement of dependence the thickness of the vapor film on the radial coordinate in the considered range of parameters. Change the drop floating conditions leads to a corresponding transformation of the interface. This may play a significant role in determining of the stability conditions for the vapor film.

References

1. Kryukov AP, Puzina YY (2009) Transfer processes at the film boiling on the downward heated surface. In: Proceedings of the 20-th international symposium on transport phenomena. CD-publication #96, Victoria
2. Kryukov A, Puzina Yu (2010) Non-equilibrium boundary effects influence on the interface surface formation. In: Proceedings of the 14th international heat transfer conference, CD-publication—IHTC14-22290, Washington
3. Labuntsov DA, Yagov VV (2000) Mechanics of two-phase systems. MEI publishing, Moscow (in Russian)
4. Ametistov EV, Klimenko VV, Pavlov YuM (1995) Boiling of cryogenic liquid. Energoatomizdat, Moscow (in Russian)
5. Stepanov EV (1991) The physical aspects of the phenomenon of vapor explosion. Moscow
6. El-Genk MS, Matthwest RB, Bankoff SG (1987) Molten fuel-coolant interaction phenomena with application to fuel safety. Prog Nucl Energy 1:151–198
7. Styricovich MA, Lamden BI, Kostanovskaya ME (1984) Heat mass transfer at the intermittent contact of the liquid drop with strongly superheated surface. High Temp 22(6):1158
8. Buevich YuA, Mankevich VN (1982) The theory of the Leidenfrost phenomenon. High Temp 20(6):1136–1144
9. Grigoriev VS, Zhilin VG, Zeigarnik YuA, Ivochkin YuP, Kubrikov KG (2007) Study of special features of development and collapse of vapor film on hemispherical surfaces. Heat Transf Res 38(5):449–460
10. Klimenko AV, Sinitsyn AG (1986) Heat transfer at the film boiling on the surface of the free swimming sphere. Trans Mosc Power Eng Inst 161:78–87 (in Russian)
11. Myshkis AD (1976) Hydromechanics of Weightlessness. Nauka, Moscow (in Russian)
12. Muratova TM, Labuntsov DA (1969) Kinetic analysis of the evaporation and condensation processes. Teplofiz Vysok Temper 7(5):959–967 (in Russian)

Chapter 6
Summary

At study of heat and mass transfer problems on penetrable interface surfaces the corresponding non-equilibrium parameters should be calculated. If the order of these parameters values is one and more then influence of non-equilibrium effects should be taken into account at problems solutions. Quantitative values of this influence depend on the problem type. Among them can be the following parameters.

1. The ratio of mass flux density j to the value of mass flux density owing to evaporation $j_E = \rho_s \sqrt{RT_s/2\pi}$, that is the dimensionless value j/j_E. The ratio of heat flux density on the interface to the value of heat flux density in free-molecular motion $q_{f-m} = 2P_S \sqrt{RT_s/2\pi}$, that is q/q_{f-m}. The transfer phenomena at low pressure as a rule are realized in these conditions: cryo-vacuum pumping processes, evaporation and condensation of liquid metals at relative low temperatures. Mass flux at evaporation in these conditions is restricted by the limiting matter possibilities because mass flux density can not be more than $\rho_s \sqrt{RT_s/2\pi}$. Hence, corresponding heat flux density should be smaller $r \cdot \rho_s \sqrt{RT_s/2\pi}$ in accordance with correlation $q = jr$, where r is the latent heat of vaporization per mass unit.

2. The correlation between kinetic relaxation time and typical time of real transfer process. Usually kinetic relaxation time is much smaller fluid dynamics time. In these conditions steady (stationary) from kinetic point of view correlations can be used. Application of such approach to the problems of film boiling gives useful information about this process.

3. The correlation between efficiencies transfer processes on the vapor liquid interface and in the liquid bulk. Usually thermal resistance of liquid is much larger the corresponding value for the interface, but for liquid metals and superfluid helium these values can be comparable and even thermal resistance of liquid can be smaller.

4. The ratio of main for corresponding problem pressure drops. For example in the problem of interface surface form determination at film boiling the comparison between hydrostatic head, surface tension drop and non-equilibrium pressure difference should be made.

A. Kryukov et al., *Non-Equilibrium Phenomena Near Vapor–Liquid Interfaces*, SpringerBriefs in Applied Sciences and Technology, DOI: 10.1007/978-3-319-00083-1_6, © The Author(s) 2013

5. The application of molecular-kinetic methods to analysis of evaporation and condensation at the presence of non-condensable gases provides these processes accurate description in limiting cases when vapor density trends to zero or unit.

Acknowledgments This work was supported by Russian Foundation for Basic Research (Project No. 11-08-00724).